Jones

The Call of
Stories

Books by

ROBERT COLES

CHILDREN OF CRISIS, I:
A Study of Courage and Fear

STILL HUNGRY IN AMERICA

THE IMAGE IS YOU

UPROOTED CHILDREN

WAGES OF NEGLECT (with Maria Piers)

DRUGS AND YOUTH
(with Joseph Brenner and Dermot Meagher)

ERIK H. ERIKSON: *The Growth of His Work*

THE MIDDLE AMERICANS (with Jon Erikson)

THE GEOGRAPHY OF FAITH (with Daniel Berrigan)

MIGRANTS, SHARECROPPERS, MOUNTAINEERS
(Volume II of Children of Crisis)

THE SOUTH GOES NORTH *(Volume III of Children of Crisis)*

FAREWELL TO THE SOUTH

A SPECTACLE UNTO THE WORLD:
The Catholic Worker Movement (with Jon Erikson)

THE OLD ONES OF NEW MEXICO (with Alex Harris)

THE BUSES ROLL (with Carol Baldwin)

THE DARKNESS AND THE LIGHT (with Doris Ulmann)

IRONY IN THE MIND'S LIFE:
Essays on Novels by James Agee, Elizabeth Bowen, and George Eliot

WILLIAM CARLOS WILLIAMS:
The Knack of Survival in America

THE MIND'S FATE:
Ways of Seeing Psychiatry and Psychoanalysis

ESKIMOS, CHICANOS, INDIANS
(Volume IV of Children of Crisis)

PRIVILEGED ONES:
The Well-off and the Rich in America
(Volume V of Children of Crisis)

A FESTERING SWEETNESS (*poems*)

THE LAST AND FIRST ESKIMOS (with Alex Harris)

WOMEN OF CRISIS, I:
Lives of Struggle and Hope (with Jane Coles)

WALKER PERCY: *An American Search*

FLANNERY O'CONNOR'S SOUTH

WOMEN OF CRISIS, II:
Lives of Work and Dreams (with Jane Coles)

DOROTHEA LANGE

THE DOCTOR STORIES OF
WILLIAM CARLOS WILLIAMS (ed.)

AGEE (with Ross Spears)

THE MORAL LIFE OF CHILDREN

THE POLITICAL LIFE OF CHILDREN

SIMONE WEIL: *A Modern Pilgrimage*

DOROTHY DAY: *A Radical Devotion*

TIMES OF SURRENDER:
Selected Essays

HARVARD DIARY:
Reflections on the Sacred and the Secular

THAT RED WHEELBARROW:
Selected Literary Essays

THE CALL OF STORIES:
Teaching and the Moral Imagination

For Children

DEAD END SCHOOL

THE GRASS PIPE

SAVING FACE

RIDING FREE

HEADSPARKS

The Call of Stories

TEACHING AND THE MORAL IMAGINATION

Robert Coles

A PETER DAVISON BOOK

Houghton Mifflin Company

BOSTON

Library of Congress Cataloging-in-Publication Data

Coles, Robert.
The call of stories : teaching and the moral
imagination / Robert Coles.

p. cm.
"A Peter Davison book."
Bibliography: p.
Includes index.
ISBN 0-395-42935-8
ISBN 0-395-52815-1 (pbk.)
1. Coles, Robert. 2. Coles, Robert — Books and reading.
3. Literature — Study and teaching — Moral and ethical aspects.
4. Education — Moral and ethical aspects. 5. Books and reading —
Moral and ethical aspects. 6. Authors, American — 20th century —
Biography. 7. Psychiatrists — United States — Biography.
8. Educators — United States — Biography. I. Title.
PS3553.O4745Z463 1989
808'.0092'4 — dc19 88-26659
CIP

Printed in the United States of America

HAD 12 11 10 9 8 7 6 5

Chapters 4 and 8 were previously published, in somewhat different form, in *Poetry* magazine, vol. 152, no. 5 (August 1988) and the *American Poetry Review*, vol. 17, no. 6 (November–December 1988), respectively.

The author is grateful for permission to quote from the following works:

William Carlos Williams, *Paterson*. Copyright © 1946, 1949 by William Carlos Williams. Reprinted by permission of New Directions Publishing Corp.

Excerpts from "Dying: An Introduction" and "A Deathplace," from *Hello, Darkness*, by L. E. Sissman. "Dying: An Introduction" copyright ©1967 by L. E. Sissman. First appeared in *The New Yorker.* "A Deathplace" copyright ©1969 by L. E. Sissman. First appeared in *Harper's Magazine*. By permission of Little, Brown and Company.

Jorie Graham, *Erosion*. Copyright ©1983 by Princeton University Press. Reprinted by permission of Princeton University Press.

To the memory
of my mother and father

Contents

Introduction

THIS BOOK BEGAN in hearing my mother and father read to each other from novels by George Eliot and Dickens and Hardy and Tolstoy during my elementary school years. My brother Bill (now a professor of English) and I had our own agenda—a host of radio programs to which we were devoted. While we responded with noisy laughs to "Easy Aces" or "Amos 'n Andy," our parents were trying to understand how England changed in the Age of Reform *(Middlemarch)* or how Russia survived Napoleon's ambitious, greedy plunge eastward *(War and Peace).* I remember one evening, when my father was saying good night to me, we had, as was our custom, what he called a "brief evening chat." Putting on a serious face and feeling a bit nervous at the prospect of challenging what by then I knew to be a much-cherished activity, I managed to voice my query: "Why do Mom and you read out loud to each other?" He wasn't surprised by the question, but he wanted to know how much thought I'd given to the subject. He asked me whether I'd discussed his reading habits, and my mother's, with anyone else. I told him that Bill and I had indeed learned that none of our neighborhood friends had parents who were so inclined. I also told him that I had tried reading a book, *Robin Hood,* with a good friend who lived next door, Benedict, and we had concluded that the exercise was slow, cumbersome, boring. Dad, sensing more at stake, wondered whether I'd be happier if he and Mom read upstairs rather than in our living room.

I recall the brief, unsuccessful fight I waged to keep a straight face. He never pushed the matter beyond that point; my smile provided him his answer: their reading had effectively denied us the use of the big parlor radio. We had to sit in the kitchen, on straight wooden chairs, listening to the small Zenith on the side table where the toaster also nested. How much more we got out of "The Shadow" on Sunday afternoons, or the programs mentioned above on weekday early evenings, after supper, when we could sprawl on the big stuffed sitting chairs, with the footstools nearby, and hear what that enormous Philco set had to offer!

I can still remember my father's words as he tried to tell me, with patient conviction, that novels contain "reservoirs of wisdom," out of which he and our mother were drinking. A visual image suddenly crossed my mind—books floating like flotsam and jetsam on Houghton's Pond, near Milton, Massachusetts, where we lived. I never told my father what had appeared to me, but he knew its essence by my glazed eyes. He made his pitch anyway: "Your mother and I feel rescued by these books. We read them gratefully. You'll also be grateful one day to the authors." *Grateful!* I was most certainly grateful, seconds later, when I felt my father's facial stubble on my cheek. No more moral explication—only that last, Yorkshire-accented "Good night, Bobby," and my ever-optimistic rejoinder: "See you in the morning."

On the way to school the next morning, my brother and I talked about "them," about their habit of reading to each other. We were no psychologists, but we were searching for an explanation that would account for motives. We found one easily: they were prejudiced against the radio (all they ever wanted to hear on it was the news), and they were trying to incline us similarly. In effect, those novels were the instruments of a parental ruse. My brother, a bit rasher than I, wanted a confrontation. I urged restraint—lest we lose what we had, that small radio's daily magic.

Not that I was averse to taking the measure of the enemy. While my brother rehearsed speeches to our parents, I surveyed their bookshelves, picked up some of their favorite novels or collections of short stories, flipped the pages with a mixture of irritation and impatience, read a few chapter titles, and brooded briefly over the propagandistic uses of these "reservoirs." I recall, once, wondering whether they would even notice if "someone" removed some of the books and hid them, or, better, threw them out.

In high school, in the classroom, I started getting reacquainted with some of those books. Yes, I thought to myself, I've not only heard of Dickens and Hawthorne, I've heard their words read *aloud.* I've heard them discussed as if they mattered more than what Fred Allen or Fibber McGee and Molly had to say. By then the English teachers were saying more or less what my parents had said, though not always, I began to notice, with the same conviction. In fact, some of those teachers seemed almost as bored with the books they assigned us as I had been. A friend of mine whose job it was to clean our English teacher's room on a particular morning (he'd been making too much noise in a study period) reported seeing a well-worn study guide of *A Tale of Two Cities;* in it were a summary of the plot and a list of all the characters. We'd been cautioned about "trots" in Latin, but no one had said anything about summaries of the novels we were reading in English. I went to the library that weekend and secured an outline of that Dickens novel, only to make the mistake of leaving it on my desk at home. My mother stumbled on it while dusting, and soon enough I was arraigned: why take such a shortcut? Dickens was meant to edify and entertain, not to become someone's "fast study," as she put it. Had our teacher suggested such a resource? No, but he'd used it himself. Well, too bad for him! That evening at dinner my parents told us about *A Tale of Two Cities,* about their appreciation of the book, their delight in it. They gave us a flesh-and-blood account of the English and French history upon which the novel drew, and we were impressed. That plot summary seemed skeletal in comparison.

By the time I went to college I shared the literary enthusiasms of my parents. They—she from Iowa, he from England—had met in Boston through a lucky accident. She loved Willa Cather, Mark Twain, Booth Tarkington, Sinclair Lewis, and, later, Wright Morris—writers who evoked the prairie states, favorably or critically or in both ways. He was a great admirer of the Victorian trio, Dickens, Eliot, Hardy—and he himself (we were often reminded) was born a week or so before long-lived Queen Victoria had died. Together my parents had read American writers such as Hemingway, Fitzgerald, and Faulkner while my brother and I were young children, and together they had also read Chekhov, Dostoevsky, and Tolstoy. Though my father did not dislike the Russians, he never took to them with the passion my mother felt. She read and reread Tolstoy all through her life —*Anna Karenina* three times, short stories such as "The Death of Ivan Ilyich" and "Master and Man" repeatedly, and *War and Peace* twice.

When I was a high schooler of sixteen and inclined to be moody and troublesome at home, my mother bought me a copy of *War and Peace* and asked me earnestly, pleadingly, to spend the summer reading it. I refused immediately, of course, asking why in the world she was pressing on me such a big book, written almost a century earlier. She answered that I might obtain a modest amount of sorely needed "understanding" (a word she favored) with Tolstoy's help. Moreover, since the Second World War had just ended and I'd been much interested in its various battles, she reminded me several times that Tolstoy knew history exceedingly well and evoked in his great novel the personal side of war—the way it gets worked into the lives of individual men and women. I took to reading the novel, but on the sly. I kept working at it for many months, even during the school year, when lots of homework had to be done. By then, needless to say, my parents were only too aware that I was ignoring math and history and Latin (and even English) for my rendezvous with the Rostovs and the Bolkonskys in a Russia being battered by Bonaparte. My mother's interest in Tolstoy's "understanding" gave way to a stubborn inability to find any understanding for me and my choices, which she deemed "out of kilter" (a serious accusation). I was "using Tolstoy in a foolish way," or so she said.

I managed to finish Tolstoy's novel by the time I finished high school, and in college I promptly enrolled in a course titled "Tolstoy and Dostoevsky"—on the naive notion that if I'd read the longest of all novels, I'd have an easy time of it in at least one freshman course. Soon I was struggling with a novel a week, many of them a mere few hundred pages shorter than *War and Peace.* I was also taking another heavy reading course—an introduction to English literature—and I began to feel myself drowning in print. For a while those "wonderful stories," as my mother sentimentally referred to them, became a torment; I yearned for a math course, or a composition class, in which I could write, rather than read and read and read in order to be able to spot passages correctly or identify in multiple-choice tests the obscure characters who come and go in long nineteenth-century novels. (My parents were shocked when they saw those identification questions on the examination in my "Tolstoy and Dostoevsky" course—an effort on the teacher's part to make sure we read the books well enough to remember the names of the characters. "Tolstoy would be horrified," my mother exclaimed. I wondered how she knew.)

By my junior year in college I had fallen under the spell of Perry Miller, who taught American literature from the angle of the Puritan tradition. He was interested in the attraction that tradition held for many writers, and in general he responded to the moral as well as to the aesthetic side of literature. But his tastes were broad and sometimes surprising. He got me interested in the poetry and prose of William Carlos Williams—hardly a Puritan divine—and eventually persuaded me to write my undergraduate thesis on the first two books of *Paterson,* published in the late 1940s. He also persuaded me (working against shyness and fearful pride) to send that thesis to Dr. Williams, who promptly wrote back with a friendly critique and an invitation to drop by if I was ever "in the neighborhood."

As a consequence of doing so, I found myself in a bind: I'd expected to get a job teaching high school English or else to do graduate studies in literature or a religious studies program (another consequence of tutorials with Perry Miller), but now I had the idea of being a doctor like this astonishing and inspiring Doc Williams, as all his patients called him. He'd taken me on his rounds, taken me to Paterson's tenements. In a way, I became one more of his admiring patients; in time I enrolled in premedical courses, went to medical school, embarked on a pediatrics residency, and eventually took training in psychiatry, child psychiatry, and psychoanalysis.

For the last decade I have taught courses for college students and those in graduate school as well. I started doing so in a modest way. I'd come back east from a two-year stint in New Mexico, where I'd gotten to know Indian and Spanish-speaking children. I was writing up the results of that study when a friend who was a Harvard administrator suggested I teach a seminar for freshmen at his college. He had in mind a course devoted to social science inquiry of the kind I had been doing in the American South and West, but I hastened to tell him I would much rather have the students read James Agee's *Let Us Now Praise Famous Men* and George Orwell's documentary writing—*Down and Out in London and Paris, The Road to Wigan Pier, Homage to Catalonia*—than any series of sociological texts. He liked the idea, and I started a seminar in the "literary documentary tradition," looking at the way novelists and poets write about certain social and political issues. That seminar has been running ever since.

In 1978 I began teaching a course—it is still a mainstay of my autumn schedule—titled "A Literature of Social Reflection." For this lecture course (with sections for discussion) we read fiction in hopes of doing moral and social inquiry. Among the works we think about are Ralph Ellison's *Invisible Man,* Tillie Olsen's stories, collected as *Tell Me a Riddle,* William Carlos Williams' *White Mule,* Flannery O'Connor's stories, Walker Percy's *The Moviegoer,* Georges Bernanos' *The Diary of a Country Priest,* Ignazio Silone's *Bread and Wine,* and some of my parents' favorites—*Middlemarch, Great Expectations, Jude the Obscure.* As in my freshman seminar, we call upon Agee, the poet and novelist, and Orwell, the novelist.

I have taught Harvard medical students in a seminar called "Literature and Medicine," where we read O'Connor, Percy, and the "doctor stories" in which William Carlos Williams gave an account of his experience as a physician working among the poor in northern New Jersey. We have also called upon Anton Chekhov, another writing physician, and Leo Tolstoy ("The Death of Ivan Ilyich"). I have found myself constantly learning new ways of interpreting those fictions—taught by my undergraduates and medical students. Differences in interpretation become apparent when I teach *The Moviegoer* or an O'Connor story such as "The Lame Shall Enter First" to college youth as against physicians-in-training. Some students who have taken both my undergraduate and my medical school courses have commented upon how a few years of life and a different intellectual agenda affected their response to a particular story. By the early 1980s I was beginning, at last, to see why my parents kept *re*reading certain favored books in the course of their lives.

As I have continued to do psychiatric work with children, I have gradually realized that my teaching has helped that work along—by reminding me how complex, ironic, ambiguous, and fateful this life can be, and that the conceptual categories I learned in psychiatry, in psychoanalysis, in social science seminars, are not the only means by which one might view the world. As I interviewed teachers during springs and summers, here and in Europe, Africa, and Latin America, I increased my autumn teaching load. I taught a course at the Harvard Graduate School of Education in which students read how teachers and their work are regarded and evoked by fiction writers—Dickens in *Hard Times,* Flannery O'Connor in "The Artificial Nigger." While studying the manner in which children's political attitudes are shaped, I gave a seminar at the Institute of Politics at Harvard's Kennedy School of Government, centered on Robert Penn Warren's *All the King's Men,* a novel that aims to comprehend what draws people into political life. Listening to aspiring teachers or governmental administrators talk about their working lives in response to the reading of one or another novel was not unlike listening to young people describe their moral and political experiences.

Later I decided to extend my teaching further, to see if novels and stories might be of use to others in other parts of Harvard University. I began a course at Harvard Law School called "Dickens and the Law"—a seminar devoted to discussion and analysis of novels such as *Bleak House, Great Expectations, A Tale of Two Cities,* and *Little Dorrit,* in which lawyers and legal questions are constantly presented to the reader. At Harvard Business School (and in an evening course for working people at the Harvard Extension School) we read fiction such as Fitzgerald's *The Great Gatsby* and *The Last Tycoon;* Saul Bellow's *Seize the Day;* John Cheever's stories; Dr. Williams' so-called Stecher trilogy; Dr. Percy's *The Moviegoer* and his essay "The Man on the Train"; Tolstoy's story "Master and Man." With Harvard Divinity School students I did readings from Percy and O'Connor, Tolstoy and Dostoevsky, Silone, Bernanos, François Mauriac—novelists who interweave spiritual matters with their storytelling. Finally, one spring, I read Henrik Ibsen's play *The Master Builder* with two Harvard Graduate School of Design students, future architects who had taken my undergraduate course.

The chapters that follow are the result of a kind of fieldwork—a teacher's conversations with his students over many years in many classrooms. The conversations have been about certain books, which I use and use and use in various courses. Put differently, the conversations have amounted to a collective exploration of the personal responses of various American students to a particular literary tradition. Many of those conversations have been tape-recorded, and here I share moments in them. They have been edited, shaped, for this book. I have attempted in these pages the same kind of documentary study or psychiatric anthropology I have presented elsewhere (in the *Children of Crisis* series, for instance), though now it is of a hometown or localist kind. Instead of venturing across regions or oceans or railroad tracks to talk with one or another "them," different by virtue of racial, cultural, or social condition, I have tried to probe my own teaching world, an intellectual and moral territory familiar to many of us twentieth-century Americans who have gone to college and perhaps graduate school.

I draw upon a number of wonderful novelists and poets in this book, but now I simply want to acknowledge an enormous debt to the students, who have taught me so much. My book's title is autobiographical: one keeps learning by teaching fiction or poetry because every reader's response to a writer's call can have its own startling, suggestive power, as my parents tried to convey. To my office hours in Adams House, at Harvard College, students brought their papers, their proposed projects, and themselves with all their diversity, perplexity, apprehensiveness, or self-confidence. Two of those students became good friends of mine and helped me a great deal—Wayne Arnold and Jay Woodruff. So, in recent years, has Phil Pulaski. I want also to mention some other very satisfying teaching I've been privileged to do: as a visiting professor of public policy at Duke University, where on numerous occasions I discussed novels with both undergraduates and medical students; as a visiting professor at the University of North Carolina at Chapel Hill; and as a visiting professor at the University of Massachusetts Medical School, where with Sandra Bertman I taught a "medical humanities" course much like the one I teach at Harvard Medical School.

My wife has been a schoolteacher for over a quarter of a century. She has taught in private schools in Massachusetts and Georgia, and in public schools in both of those states and in Louisiana and New Mexico, where we have lived while doing our work together. Her subjects have been English and history, and her high school children have come from a wide range of social, racial, and economic backgrounds. Since our marriage she has more than reinforced my attachment to novels, stories, poems, and plays. I have often acknowledged her decisive influence on me. I cannot imagine myself, without her, doing the work I have done. Peter Davison, who has edited my writing ever since it began, is a fine poet, a good friend, a person of great breadth and depth, and, to me, a wonderfully energetic and knowing teacher. Like my wife, he has taught me by example. I thank him, as well as my wife, even as I salute the lives of my mother and father, both of whom, in their eighties, died near the end of 1985 after sixty years of a most gracious and loving marriage. God bless their memory.

The Call of Stories

· I ·

Stories and Theories

A TALL, THIN, WAN LADY with light skin, wide blue eyes, and black hair, she had a distinct presence on the psychiatric ward of the Massachusetts General Hospital, an institution otherwise devoted to medicine and surgery, where I began work early in July 1956. I remember noticing her the first time I entered the ward. She was not walking but pacing, covering an area of the hall she had circumscribed for herself. Her vigor and tenacity made their mark on everyone else: even the doctors kept out of her way. Every time the door to the locked ward opened, however, she stopped for a second or two, looked at the person entering or leaving, waited until the noticeable slam had taken place, and then resumed her movement. Sometimes it was a relaxed stroll, but mostly she was marching herself at the behest of orders no one had managed to fathom.

For the first two days we residents attended meetings and orientation lectures; on the third day we were given names and told to go seek the persons who bore them, whereupon those persons would be regarded, in the community known as a psychiatric service, as our patients; and we, of course, would be their doctors. I still remember the moment when the names, handed to me on pieces of paper, became reality—eight men and women in all. For a start. We would "pick up more," we were told, when we did our weekly stint of emergency ward work: "All those you admit are yours." I still remember reading those names on the pink slips of paper, then setting out to find "them"—the people I would be "treating."

The black-haired woman was the first person I met on my search. Another resident had told me—warned me—that I was getting "the hiker." "Will you ever get her to sit and talk?" he asked skeptically. None of us, up till then, had seen her in a chair. She ate her food upright, keeping her feet moving, always moving.

I was frightened—not only of her driven activity, of the tension that seized her face when the door opened, of the furtive glances she gave anyone who came near her, but because I well knew my own ignorance, my inexperience as a doctor. Still, we'd been told that the patients were all anxious and were eagerly, worriedly, awaiting their new doctors. Best to get started immediately, they'd said, lest already troubled people become even more so. So I hastened to approach the ward's constant walker. Soon enough I too was walking while I told her who I was and asked if we might at some point have a conversation. She looked beyond me, toward the door, actually, as she said yes, but she pointed out that we were having a conversation right then. She'd be glad to continue it. Was I sure I wanted to talk with her? I didn't have it in my heart to say yes. I nodded, though. The truth was that I sensed how hard it would be for me to do much for her, even to earn a preliminary trust from her. Nor did I enjoy the prospect of doing psychotherapy out in the open, in full view—up and down, up and down. "She stops to watch television occasionally," a nurse told me, and added: "Maybe she'll stop to talk with you."

I was not hopeful after our first encounter. We walked back and forth, our exchanges brief and not at all encouraging. I tried to get her to answer my open-ended questions ("Would you want to tell me what brought you here?"), but she would have no part of them. "I just came," she said. I can remember what crossed my mind next: What brought *you*, the doctor, here?

In time, with more self-confidence, I would not need to ask myself such questions, at least not in the course of a psychiatric interview. But that day I gave up fast, promised the walking woman a return, and went on to my next patient, a man all too ready and willing to talk. I'd been told he was "hypomanic," and in a few minutes I learned the practical significance of such a characterization—an unrelenting volubility. I found my mind wandering back to the person I'd seen earlier. Why wouldn't she talk just a bit more? As the second patient accelerated his output of words and loosened the logic that supposedly connected his sentences, I asked myself a companion question: Why did he talk so much—and wouldn't it be great if his "hypomanic" behavior were as contagious as some authorities claimed can be the case? *He* could stir things up a bit in *her* and thereby, incidentally, help *me* out.

That week, I had to bring what were called "protocols" to the offices of my supervisors. When I went for the first time to see the two senior psychoanalysts who were to help me make sense of what I was hearing, I was quite nervous—not sure what to say, how to say it. I'd done some walking myself before going to each of the two offices, and when I sat before each of those doctors, those middle-aged men, I was torn between two inclinations: to say very little, if anything, lest I reveal my stupidity, my inadequacy, and maybe my "problems" (a word constantly in use by everyone—doctors, nurses, patients—on the ward), and to keep talking at a brisk pace, lest awkward silences develop, a clue no doubt to—well, my stupidity, my inadequacy, and maybe my "problems." I recall noticing that one of the doctors, Carl Binger, had bifocal glasses; and that the other one, Alfred O. Ludwig, cupped his ear repeatedly, at which point I could feel my vocal cords tighten, my voice get louder. I recall, most of all, the assistance those two doctors offered me.

In our first meeting Dr. Binger urged me to read more in the psychiatric literature so that I might understand "the nature of phobias." He could see that the suggestion puzzled me. "She is phobic, and you've got to work around her defenses." It was clear, however, that I hadn't a clue as to how I might follow his advice. "Best to see her in your office. Have the nurse bring her there." A pause, and then an explanation: "Phobics are power-conscious. You'd do well to make it clear that you're the doctor, and that you intend to see her in your office at a time of your choosing."

He was, of course, trying to strengthen the resolve of a fledgling, an apprentice in psychiatry, who wasn't at all sure he *was* the doctor and who hadn't yet become as time-conscious as one becomes when in practice. He was giving me words to grasp; he was "treating" the floundering "doctor" so that the doctor could in turn "treat" the "patient." Words like "treat" and "patient" provided considerable support to me when moments of self-doubt arrived, as they often did. Add "phobic" and "defenses," add the suggestion of "working around" the latter, and—presto!—one has a stated enemy, a military strategy, a purpose.

I set to work. I mobilized my authority as I'd already learned comfortably to do—by telling a nurse what should be done. I noticed her look, for a long second, right into my eyes. But I was busy, and I was now in possession of knowledge, maybe even wisdom: Dr. Binger, with distinctions galore, had given me an explanation, a suggestion. What *else* was there to do—bow endlessly to a string of compulsions? (The walker was a demanding, finicky eater; she constantly turned down trays of food, asking for substitutes not so conveniently secured from the hospital's kitchen.)

Meanwhile, I had the second supervisor to meet with. In my mind Dr. Ludwig became, for a while, the man who made me speak up, speak loud. My own phobias had always urged upon me a policy of withdrawal, escape, silence, or, at best, brief and softly spoken comment. That was the way I handled professors whose prominence and outspokenness intimidated me, made me feel, in fact, speechless: no words of mine could have any value to people who had used words as they did, thereby becoming the important people they were. This doctor, though, seemed desirous of hearing me out, and if I faltered and fell silent, or if my voice fell a bit, he moved his right hand to his right ear. But he himself had little to say. He did, at least, I realized after a few meetings, get me going.

Dr. Binger was fast with the lectures; Dr. Ludwig, the man with the flawed hearing, worked our time together in such a way that he really had to use his hearing long and hard. With Dr. Ludwig I found myself becoming increasingly, relaxedly discursive; with Dr. Binger, I was ever fast to offer conceptualizations. He wanted them, too. "Let's try to formulate this case," Dr. Binger would exhort, and if I hesitated, he was ready to weigh in, much to my pleasure and edification. He was known as a brilliant theorist, and he saw me as the lucky resident who, early in his career, was learning to follow suit. Dr. Ludwig was known as a nice guy; he was also regarded as a bit slow on the draw, perhaps over the hill. His hearing trouble was not a consequence of old age, we'd heard, but rather of some neurological disorder. Nevertheless, we regarded him as an affable old gent. (I now realize he was in his mid-fifties at that time, my age as I write these words.)

I began to realize, a month into that residency, that I wasn't getting very far with the "phobic" patient. The nurse had indeed managed rather decisively to persuade her to come and see me in my office, to sit down while there. I had learned to move past some of her "defenses" all right, disarm her with my slyly abrupt and probing questions, many of them borrowed from my supervisor—for example: "Does the walking help you with all that anger?" This was put to her at a time when we'd never once discussed her anger or anyone else's. She looked at me blankly for a second, and I got apprehensive. (Could I have been angry?) But she broke into a smile, said she was glad I was her doctor, because I'd "caught on" to something that had been bothering her for a long time, her anger. I was stunned, and began to feel quite pleased with myself. I recall sitting back, stretching a little (enough for her to take notice), and thinking to myself: you *can* effect change in this field of psychiatry—if you know what you're doing. Still, the patient's general behavior in the ward (and toward her husband, whom she mostly shunned) did not change, and a vigorously analytic Dr. Binger reminded me over and over that "phobics are hard to treat," even though we can ascertain so much about their "psychodynamics," a word he constantly used: "What are the psychodynamics at work here?" (Once I looked around the room, forgetting that by "here" he meant the patient's mind.)

In another part of the hospital (or forest!) I was still trying to make sense of that same patient's difficulties with the help of my hearing-impaired adviser. One early morning—Dr. Ludwig saw me at eight, before he saw his patients, and there were days when I felt like one of them—I had very little to report, so I felt apprehensive: all those minutes, with at best contrived talk. The doctor across the room took quick stock of the situation and told me *he* wanted a little time that morning, if it was all right with me. He announced that he was going to tell me a "story." My ears perked up. I recalled my father sitting on a chair in my bedroom, telling me stories before I fell asleep. Dr. Ludwig's story concerned a patient, a woman almost paralyzed by various worries and fears. The doctor told me a very great deal about her—where she'd grown up, her schooling, her hobbies and interests, the reading she did, the programs she watched on television, the clothes she bought, even where she bought them, and most of all, the *events* in her life: where she met her husband and how, where she traveled and why, where she spent her spare time and with whom. I was quite taken up by listening, even forgetting for a long spell that this was a patient's "clinical history" I was hearing.

Suddenly the story stopped: the patient had been struck by a car, on the way to a lecture at an art museum. I was surprised, saddened. I felt questions welling up in me. What happened to her as a result of the injuries she sustained, and in general, as she got older? She had a name, and Dr. Ludwig had been using it; and I was using it, too, as I pictured her in my mind—saw her being hit by the car, taken by ambulance to a nearby hospital. Dr. Ludwig suddenly stopped to think, though; he sat and looked at me. I wondered why, what to say, to ask. The silence was broken by his question: "Do you see her in your mind?" "Yes," I answered. "Good," he responded.

"I have told you a story," the doctor said. Nothing more. I awaited an amplification in vain. It was my turn. I responded to the storyteller, not the doctor, the psychiatrist, the supervisor: "What happened?" I was a little embarrassed at the sound in my own ears of those two words, for I felt I ought to have asked a shrewd psychological question. But Dr. Ludwig said he was glad I'd asked the question I did. Then he told me "what happened." Afterward there was a different kind of silence in the room, for I was thinking about what I'd heard, and he was remembering what he had experienced. Finally he gave me a brief lecture that I would hear in my head many times over the next three decades: "The people who come to see us bring us their stories. They hope they tell them well enough so that we understand the truth of their lives. They hope we know how to interpret their stories correctly. We have to remember that what we hear is *their story.*" He stopped there, waited for me to speak. But I had nothing to say. I hadn't quite thought of my patients as storytellers and was letting that settle into my mind. He started in again, now more expansively and didactically. He reminded me that psychiatrists often hover over their patients, intent on "getting a fix" on them: make a diagnosis; ascertain what "factors" or "variables" have been at work; decide upon a "therapeutic agenda." He wasn't criticizing such routine evaluative procedures, nor did he have any dramatic alternative to them. He simply wanted to remind me that I was hearing stories all day long, and that when I came to him for supervision I was bringing stories to him—telling a story at second hand.

What Dr. Ludwig said was pure common sense, yet it gave me a jolt. Why? Much of my mind's energy was then taken up with abstractions. I'd been acquiring facility with abstractions in one school setting after another—an effort that is, inevitably, both self-fulfilling and self-serving: the rise of someone through the junior ranks of the academy. But now a difference obtained: I was learning new abstractions and using them not merely to impress section men or women in courses, or professors during their office hours, or myself while brushing my teeth and musing over the new day's opportunities, but to help understand the palpable pain and suffering of another human being. Although my sessions with one supervisor or the other had the old academic flavor of an initiate eagerly trying to demonstrate his growing capability with certain ideas, there was a third person in those consulting rooms (or should have been), and that person's life was being turned into a "text." It was to this aspect of supervision that Dr. Ludwig was referring that day.

He pressed the matter further, actually, as we both got up to say good-bye. Indeed, I would notice over the months that his more trenchant and, for me, lasting comments often came just as we were ending our time together, almost as if he wanted me to think about something in my own good time, and maybe wanted to remind himself of what he, too, might occasionally forget. "Supervision" is after all a meeting of two persons, a shared possibility for each of them. As I moved toward the door, Dr. Ludwig made a suggestion: "Next time let's talk about some events in her life; for a while we can put aside formulating her problem."

I took the suggestion as a criticism. Nervously insecure in those days, I took many suggestions as muted or all too explicit judgments on what I was doing wrong. Meanwhile, Dr. Binger kept encouraging me to "formulate the problems" I was trying to treat. With regard to that phobic patient, he had suggested a "reformulation"—a new "therapeutic strategy" based on a new appraisal of "the state of her psychodynamics."

I found him, I have to admit, a very helpful person to go see. He talked a lot. I think, in retrospect, I got an intellectual fix from him. I came into his windowless, darkly lit office, sat where his nonanalytic patients did, delivered my prepared speech, reading aloud notes I'd taken while the patient spoke, or thoughts set down afterward, and then waited while a learned, marvelously articulate and self-assured psychiatrist told me what was "really" happening in those "therapeutic sessions." When he had finished his interpretative foray, assessed the state of the "transference" (the phobic patient's responses to me, based on her past experiences), and given me a brief lecture on the psychoanalytic theory of phobia formation, I often found myself feeling less afraid for myself, if not for the patient. I now "knew" her, and I could look forward to yet another chance to listen, to inquire, to hear confirmed what I'd been taught. As for the occasional moments of doubt or worry (Why isn't she getting any better?), my supervisor had some analgesic words: "These are difficult problems to treat, phobias. Your job is to understand her and help her understand herself. She is a rigid personality with a serious character disorder. Even with years of psychoanalysis her defenses would be hard to undo. Try to learn, and if she can use you to her advantage, profit from the relationship and the insight you offer, well and good. But concentrate on understanding her, not on trying to change her behavior."

He was, I still believe, quite correct in some important respects. Years of clinical experience had most certainly taught him the intensely refractory quality in "symptomatology," a word he favored. Moreover, I was an apprentice whose long years of education actually gave me little help in this situation, wherein I sat with someone who was acting very strangely and tried to learn more and more about what was happening in her mind: her thoughts, worries, dreams, and quite a number of nightmares. (When I heard her accounts of nightmares, I could feel hopeful about my career, if not the patient's future, for I could anticipate my supervisor's concentrated attention upon those middle-of-the-night perturbations, the "pay dirt," as he once called them. I learned it was best for me to be "cautious, politic, meticulous"; best for me to set aside therapeutic zeal; best for me not to get "too involved." That phrase was a mainstay of our daily intercourse as residents. We were echoing our supervisors' advice in aiming at that "free-floating attention," that "hovering distance" whose fulfillment would mean the achievement of professional maturity and competence. Of course, those phrases contained a touch of poetry, a bit of implied magic—but we weren't about to look at them in so literary a way. Nor were they given to us as words of inspiration; rather, they were prescriptive, summing up a "position" so firmly that when the phrases were uttered by someone (or by myself, to myself), I could feel my leg muscles tighten a bit, or my eye muscles: getting set to put myself in proper place, as it were.

Yet "that other supervisor," as I began to call him in my mind (the "other" signifying for me a certain gradual divergence on his part from prevailing practice), kept nudging me in a different direction. Eventually I heard myself one morning, to my own surprise, tell my phobic patient, in a moment of frustration (we weren't getting anywhere, we were covering the same material again), that I wanted to hear more about *her,* not about the "symptoms" she had learned so well to describe to me and to the nurses, social workers, occupational therapists, physical therapists, ward secretary, group therapists, and other patients. What did I mean? I wasn't sure how to answer her question, which she quite predictably put to me. But now that she asked, I found myself prepared to spell out answers rather vigorously. I said that we had spent a good deal of time discussing her various fears and how she tried to come to terms with them; now it would be a good idea for us to pay sustained attention to her *life,* to its course over a span of some thirty-five years. She responded with another question, spoken in tones of irritation, confusion: hadn't we been doing that all along, hadn't she told me about her life, hadn't I taken her history?

Oh, yes, I had taken her "personal history," her "family history," her "social history," her "clinical history"—all those phrases that were established in hospital residents' minds and formed distinct elements in the bureaucracy's code of procedures. Each patient's chart, that is, had those phrases printed on separate pieces of paper, and it was our task, as house officers, to take those different kinds of histories and write them up conscientiously. But I had something else in mind, though I wasn't sure exactly what, as she pushed me with her somewhat plaintive requests. I recall saying, finally, to my own barely suppressed annoyance—perhaps it was directed as much at myself as at her—that we should try to "begin at the beginning again." To amplify that spectacularly original suggestion, I asked her to try to tell me a few stories about her life—"moments in it," I went on to say, "you remember as important, as happy or sad." Then I sat back and waited. Here you are, I was silently saying to the patient; now whatever you do with it—through your words or indeed your silence—can only be of interest, and, further, will be "revealing" (another buzz word of the times).

She was utterly quiet long enough for me to get anxious. Should I say something? Had we reached an impasse? Should I try to break it? Rewording one's questions or interpretations is a handy way of giving the patient time to think, then speak; so a host of teachers had reminded us. I was drawing in breath, bracing myself for such a tactic, when my patient, in a flat, scarcely audible voice, told me she'd had only one or two happy moments in her entire life. She had chosen to respond to those two words, "happy moments," and not to the other words I'd used. She looked to me for a signal. She was still hesitant to begin, so I said it—said what no one had suggested I say, not even Dr. Ludwig: "Why don't you just tell me a story or two?"

She looked at me as if I'd taken leave of my senses. I began to think I had: this was no way to put the request I had in mind. Why *had* I phrased my suggestion that way? I explained that we all had accumulated stories in our lives, that each of us had a history of such stories, that no one's stories are quite like anyone else's, and that we could, after a fashion, become our own appreciative and comprehending critics by learning to pull together the various incidents in our lives in such a way that they do, in fact, become an old-fashioned story. I am now paraphrasing myself in somewhat formal language, but the last phrase is word-for-word what came out of my mouth. I recall being embarrassed that I'd used the expression "old-fashioned," surprised that I'd repeated the word "story."

For the first time in my short career in psychiatry I saw a

noticeable and somewhat dramatic change take place in a patient—and not in response to any interpretation or clarification of mine, but merely as a result of a procedural suggestion, as it were: how we might get on, the patient and I. She smiled at me and asked if she could go get her cigarettes (she'd never smoked during our meetings, though on her ward promenades she became a smokestack), and then, having returned with a package of Lucky Strikes, she started off with no prodding from me. Her story poured from her: a childhood in hospitals because of a birth defect (harelip) that needed repeated bouts of reconstructive surgery; an adolescence of abuse at the hands of an alcoholic father; a brush with death, when her mother and she, speeding away from their home, where her father was violently drunk, got into a serious automobile accident; and finally, her marriage to a man who turned out to have a serious criminal record (delinquency as a youth, thievery as an adult). She had previously told us (the nurses and social workers as well as me) practically none of the above. She had presented her father and mother as quiet, friendly, "average" people who had "no problems, none really." She had told us very little about her husband other than that he started making her "nervous" about the time her agitated fearfulness began, two years prior to this hospitalization. The sexual side of her trouble was obvious to me (and more so to my supervisors), but she had not been willing to talk about such a personal matter, and both supervisors had suggested I wait a while before bringing up the subject. Now she had revealed other reasons to be afraid, and they were quite real: her husband was connected to a gang she described as part of the "underworld," and she had, without his knowing, read a death threat he had received by mail.

All this and more I heard unfold over several meetings, in a veritable outpouring. I sat there listening and abandoned the ideal of a formal, psychiatric inquiry—if I hadn't, I'd have repeatedly interrupted her, and, no doubt, stopped her in her tracks or diverted her into paths that might have brought her back to her silent wariness. At one point she asked me if she was "making sense." I put the question back to her, asked her what she meant. She said that she was telling me about her life, but she was afraid she might be wasting my time. What made her worry that that might be happening? She replied that by talking as we just had, we had stopped discussing her phobias. I pointed out to her that on the contrary she had told me about a phobia we hadn't ever examined—she'd never before mentioned how fearful she was for her husband's life. She nodded, took a lit cigarette that had been barely smoked and used it to light another one, put both in a nearby ashtray, and continued telling me about her life and the lives of her mother, her father, her husband, and her brother-in-law (he was a math teacher in a suburban high school, and she "secretly loved" him)—all the while "dragging," as she described it, on those two cigarettes, first one, then the other, in a carefully controlled alternation.

At the end of that talk she had used up not only those two cigarettes but their two successors, had joked with me about her "crazy way" of smoking ("I'm even afraid when I smoke—that I'll need a drag and I won't be able to get one, and there'll be no matches because I've lost them or someone's taken them"). She wanted to know, as she left, if I'd changed "techniques" on her. What did she mean by that question? She meant that I'd taken to "talk about life," not questions about her "problems." Might we have, thereby, wasted time? No, I insisted, we'd done just the opposite, put our recent hours to good use.

I wasn't so sure, though. I began to think I'd been deflected, either by her unwitting design or by my own inadequacy or both. Maybe I was simply committing the sort of error all first-year residents make, an error one older psychiatrist had in fact described: I'd allowed the patient's "agenda" to take over, had become bogged in a mass of circumstantial detail. Where would I go now, anyway, in the further sessions that awaited both the patient and me? I wasn't even sure I'd have time during my next supervisory hour with Dr. Ludwig to tell him all I'd heard!

I did manage to tell it all, however, and his response marked an altogether crucial moment in my postgraduate education. He pointed out to me that our patients all too often come to us with preconceived notions of what matters, what doesn't matter, what should be stressed, what should be overlooked, just as we come with our own lines of inquiry. He pointed out that patients shape their accounts accordingly, even as we shape what we have heard into our own version of someone's troubles, the "presenting history." Too often, he said, we are bent on an "abstract." That is what we are trained to do—get to the concise, penetrating heart of things, then try to help someone change. Yet such an approach, even when pursued by the most experienced clinician, can come to naught or miss too much. In explanation, he spoke these unforgettable words: "Each patient will tell you a different story, and you're an all-day listener."

He went much further, suggested I write my notes differently. In what way? I ought to write brief biographies of the patients rather than come to his office with a list of the patients' complaints. I also ought to make a list of "interesting clinical moments" in the interviews—remarks that I deemed important. To what purpose? He wasn't specific or conveniently certain in reply; he simply told me that "something happened" when we had encouraged our patient to tell a story or two about her life, and "we ought keep going in that direction," though he was quick to add, "not too vigorously or in too organized a way."

I now realize that an older doctor was trying to get a younger one to stop forcing his patients (not to mention his own mind) into a variety of theoretical constructs. He wanted me to conduct an open-ended kind of interview. He wanted me to hold off the rush to interpretation, to restrain myself from trying to get him to give me *his* interpretation, which for me would become the definitive one, at least until the next supervisor came along. He encouraged a gentler tone, a slower pace, a different use of the mind.

The upshot of that weekly experience with Dr. Ludwig was a turnabout in my way of working. Prodded repeatedly by him, I gradually found myself more interested in the concrete details of a given person's narrative than in aggressively formulating her or his "problems." I began meeting a second time each week with

Dr. Ludwig to discuss the more general problems of psychoanalytic therapy. Once, in a playful mood, yet utterly serious, this new friend and adviser cautioned me: "Remember, what you are hearing [from the patient] is to some considerable extent a function of *you,* hearing."

A year later, sitting with a young man who had attempted suicide (sleeping pills) and failed, I made the acquaintance of another "variable": the many stories we have and the different ways we can find to give those stories expression. This youth, a college sophomore majoring in English literature, was much interested in nineteenth-century poets such as Keats, Browning, Shelley, Wordsworth, Tennyson. His older sister and his mother were writers, the mother a poet, the sister an advertising copy editor by day and a graduate student, through seminars and extension courses, in the evening. The father, a scientist, had died of cancer in his forties—a landmark event for this young man. I remember quite well the challenge he put to me during our very first conversation, which took place in the emergency ward of the hospital where I worked. When I arrived, he was pale and grim, but not reluctant to talk. He told me, willingly and with some fluency, what had happened. Then he asked me—I'd been with him only an hour or so—whether there were any "woman psychiatrists" in the hospital. I was surprised at the question. I said yes, there was one woman psychiatrist on our house staff. He was quick to request that she be his doctor.

I was ready to assent on the spot, but I said no more than that I'd tell the staff of his request. "All right," he said, shrugging, "what will be will be." But as I was half out the door of the emergency room cubicle where he lay on a movable stretcher, his head propped on two pillows, his right arm by his side, with an intravenous dextrose-and-water solution flowing into a prominent vein on the back of his hand, he gave me something to think about: "I don't know if I could tell the whole story to a man; I'd tell him a different story, I'm sure."

Dr. Ludwig was quite interested in that comment, and we both became grateful to the youth, whom I *did* end up working with.

The one woman resident psychiatrist was already overwhelmed with work, and, as I've mentioned, those of us on duty acquired the patients we'd seen the previous night or weekend. Besides, the attending psychiatrist felt that in a way the young man was challenging us—maybe even asking unwittingly for what he seemed averse to having, a man he might talk with. "If he brings up the issue of the sex of his psychiatrist, talk the matter over with him," I was told.

When I went to see my new patient, he reminded me of our earlier exchange in another part of the hospital. I told him I'd done my best. He looked at me with a mixture of disappointment and barely concealed irritation, and said nothing. I began, asked those usual, apparently innocuous questions that were intended to get us going as talker (him) and listener (me). But he wouldn't go along; he started asking me, with noticeable persistence, about my life. Where had I gone to college, to medical school? What had been my major as an undergraduate? Where was I born? Were my parents still alive? What was my father's occupation? What were the origins of my interest in psychiatry? That last question did it—got me to clam up and start mobilizing *my* defenses. I'd been giving him answers to his rapid-fire inquiries because I wanted to put him at ease and be the outgoing and friendly person he seemed badly to want and need. But that last request for information went too far. Though other patients inevitably had asked it of me, none had done so after so brief an interaction and (I felt) with such presumptuousness. I had scarcely asked him anything —simply to tell me about himself.

But he clearly regarded that small request of mine as itself presumptuous. He told me that I was older than he, and acting the part, but he insisted on his "rights." Before I had a chance to explore *that* (what rights did he have in mind?), he headed me directly off: "You tell me your story, and I'll tell you mine." I fear I only got worse: I reminded him, with a tact and control meant to conceal my annoyance, that *he* had come into the emergency ward, had almost died, and that it was best that "we" do what "we" could for him as the patient—whereupon he reminded me that he had requested a woman psychiatrist. I asked him, quickly, what he'd tell her that he'd not tell a man. He wasn't about to yield to that maneuver. He stopped talking.

I decided to break the impasse: I told him about my early interest in literature, my later interest in pediatrics, my recent one in psychiatry; and I gave him some idea of how those interests had emerged from the life I'd lived. When I was through with my story, which took five minutes or so, he promptly began his. He let me know how much he had yearned to talk with his late father, which was now obviously impossible to do, and how much he had talked with his mother and sister, both of whom doted on him and invited him to talk endlessly about himself, his interests and activities. Now he'd done something that he didn't want to discuss with them. As I eventually learned, just before he took the overdose of pills he had summoned a fantasy for himself—that his father was there in the room and that they were talking and talking. He was telling his father about his life; his father was listening, was sharing with the young man a report or two about his own.

When I put that youth's "history" into a theoretical formulation, the familiar phrases appeared, none of them surprising, each of them applicable not only to that person but to many, many others: "domineering mother" (and sister), "poor masculine identification," "aloof father figure," and so on. When I named his "defenses," his "hostility," the kind of "transference" (involvement with me) he'd make, I was again consigning him (and me) to territory populated by many others. No wonder so many psychiatric reports sound banal: in each one the details of an individual life are buried under the professional jargon. We residents were learning to summon up such abstractions within minutes of seeing a patient; we directed our questions so neatly that the answers triggered the confirmatory conceptualization in our heads: a phobic, a depressive, an acting-out disorder, an identity problem, a hysterical personality.

Some of those labels or categories of analysis are psychological shortcuts and don't necessarily mean offense to patients or diminishment of the user. On the other hand, the story of some of us who become owners of a professional power and a professional vocabulary is the familiar one of moral thoughtlessness. We brandish our authority in a ceaseless effort to reassure ourselves about our importance, and we forget to look at our own warts and blemishes, so busy are we cataloguing those in others. That young man was sensitive, indeed, to the question of power—to the way it was used in a family, *his* family. He'd often wondered, I would discover, what life might have been like for him if his father had been, as he repeatedly put it, "stronger." Meanwhile he knew how to contend with his mother's "strength," and his sister's—at a cost, of course. He was afraid to take the risk of having a male doctor, who would find out, he feared, even more about him than his busybody mother and sister knew. No wonder he wanted my story to precede his—and had I been more comfortable with the give-and-take of storytelling and less inclined to coax "information" from him so that I could make a "diagnosis," I might have spared both him and me a long bout of sparring. He kept himself aloof, his story unspoken, while he tested, with his questions, my willingness to surrender enough of myself (my story) to show reasonable good will toward him.

Working with that youth, as with "the hiker," was not easy, but I learned a great deal. Dr. Ludwig kept pointing out that I was hearing complex, somewhat repetitive, but also quite pointed stories, and that my remarks were a sympathetic critic's response to those stories. Both those patients, he pointed out, had every objective reason to be afraid of me. They both had recently felt put upon, manipulated, even betrayed. They were not "crazy"; rather, they were hurt individuals who had a lot of trouble to contend with. They weren't doing very well. One had become paralyzed by her fears; the other had tried to end his fears by suicide. Each was as guarded as could be, and clearly uninterested in responding to the intense questioning of psychiatric residents. Our questioning, Dr. Ludwig pointed out to me, had its own unacknowledged story to tell—about the way we looked at lives, which matters we chose to emphasize, which details we considered important, the imagery we used as we made our interpretations. If our job was to help our patients understand what they had experienced by getting them to tell their stories, our job was also to realize that as active listeners we give shape to what we hear, make over their stories into something of our own.

Even though Dr. Ludwig helped me considerably with those two patients, and with others, I was not altogether comfortable with his way of seeing and putting things. Nor was I encouraged to be so by the other psychiatrists who supervised my work, or indeed by their professionalized world. A psychiatric assertion, "The patient is phobic," lingered in my mind as I saw the black-haired woman—as did phrases applied to the suicidal youth, such as his "borderline personality." Those words were written on an important paper known as a "chart." The writers were "supervisors," committing their knowledge and authority. They, too, fellow human beings, were of course wise one minute but fallible and maybe foolish the next, though I did not so readily appreciate it back then as I was struggling to learn. I had already been taught the comforts and rewards that come with the embrace of certain magical verbal constructs.

No wonder I was jolted one day when the one supervisor who had a sense of humor about our psychiatric enterprise, and who was constantly urging me to listen to the stories of my patients with a minimum of conceptual static in my head, announced to me that a particular patient was afraid of being "done in" by me, the way she'd been "done in" by her father. What could Dr. Ludwig possibly mean? A judgment of my inadequacy, or, worse, my poor intentions? A comment on the patient's problems—her attribution to me ("projection") of certain qualities her father possessed? A difficulty in the "transference" and maybe the "countertransference" (my response to her, a result of my own past experiences)?

Dr. Ludwig smiled broadly, which always gave me a clue that we weren't going to be immersed exclusively in psychiatric constraints. Then he began to teach, through friendly but pointed talk. He reminded me how scary it can be for a suffering person to sit and speak in the expectation that the response will be a categorical word, a diagnosis. My patient had kept asking me what her diagnosis was. I'd dodged the questions, but halfheartedly. I *was* in pursuit of *le mot juste.* I'd been handed a manual with lots of classificatory schemes. I'd noticed how confident doctors became as they got familiar with those schemes. Moreover, when supervisors discussed the prolonged relationship between a psychiatrist and a patient, the talk turned on such abstractions as "therapeutic regression" and "transference neurosis." The old sage across the room from me, putting his right hand through his unkempt gray hair, asked me to stop and weigh the gains and losses, stop and consider ways of reducing the losses. I could not forget his question of me, partly rhetorical and partly an urging: "Why not let her story keep unfolding—that and no more for a while?"

In those days I was always mentioning someone's theory, or someone else's, and so were many of our teachers. Soon Dr. Ludwig asked whether I'd given thought to "the meaning of theory." Jumping ahead of the questioner, I interpreted the question psychologically. Yes—that, too—but Ludwig first wanted to examine with me the literal meaning of the word "theory," and only thereafter its emotional significance. I hadn't the slightest idea of the derivation of the word, but he had in his office a wondrously large dictionary that he kept on a special area of his desk, as if on hallowed ground. He flipped through the pages and soon was reading me the dictionary's words. I learned that the critical root is the Greek θεάομαι, *theamai,* "I behold," as in what we see when we go to the theater. We hold something visual in our minds; presumably, the theory is an enlargement of observation.

All too commonly, however, some of us use theory more as a badge of membership than as a visual stimulus. We might use ideas in such a manner as to make us eagle-eyed; instead, often enough, we see as sheep do. Moreover, we keep saying certain words out loud, as tokens of loyalty in the company of our colleagues, and repeat them silently while we hear patients tell us not only what has brought them to us, but what they hope to gain from us. Under such circumstances, they may get what they had not bargained for—an indoctrination. As for us, the doctors, we get further confirmation of the correctness of what we resolutely believe. In the cautionary words of Dr. Ludwig, spoken at the end of a session that had my teeth chattering: "The patients are often quite sensitive to what we want of them, and when they use our favorite phrases, they are trying to show us how hard they are listening, how eager they are to please." So what? I remember thinking. Why should I find that outcome in some way reprehensible? He was turning into a crank, I decided. The antitheorist was making me even more self-conscious by warning me that in our self-consciousness as psychiatric theorists we lose sight of human particularity.

I kept my mouth shut, however. Finally he spoke: "I've heard you say several times that 'the patient feels he's regressing in the hospital'; and several times that 'the patient realizes he's going through an identity crisis.' " (We were talking about the young man who had attempted suicide.) I nodded. He reminded me that my technical language had now become my usable property (talk about a version of transference!), and not necessarily to my patient's benefit, or, for that matter, to mine as a doctor. He insisted that in and of itself such a development was not necessarily harmful; but there *were* dangers, he allowed, in a lowered voice. A pause as I wondered how broad his indictment would be. Then his soliloquy—and how sad that it was not recorded.

Dr. Ludwig wanted me to worry about messages omitted, yarns gone untold, details brushed aside altogether, in the rush to come to a conclusion. "You were right to suspect within two minutes of your first meeting with him that he was going through an identity crisis, but when you hear *him* telling *you* he's going through an identity crisis, you might stop and wonder what else he's now going through—and you might begin to worry about what he *isn't* telling you, now that he's telling you about his identity crisis."

Such talk was somewhat contrived, I thought, and I noted as

much in the write-up I did after our hour was over. (He had encouraged me to sit down and write a summary of what had transpired between him and me in those supervisory meetings.) But on that fast-darkening winter afternoon, I was urged to let each patient be a teacher: "Hearing themselves teach you, through their narration, the patients will learn the lessons a good instructor learns only when he becomes a willing student, eager to be taught."

He was, I concluded, old and venerable. Yet no one else was preaching that line of psychological inquiry. To be sure, other supervisors and lecturers were making their own ideas heard. No wonder my patients, upon hearing those ideas expressed (as they did when they eavesdropped on our doctoral conversations), took pains to join ranks with those in command. It didn't take more than a week or two for the favorite themes of a young attending psychiatrist to work their way down to the patients. I now realize that my supervisor was actually arguing for a revolution—that the lower orders be the ones whose every word really *mattered*, whose meaning be upheld as interesting. We had to change our use of the very word "interesting": no longer were we to appropriate it for ourselves. What ought to be interesting, Dr. Ludwig kept insisting, is the unfolding of a lived life rather than the confirmation such a chronicle provides for some theory.

I slowly began to realize that we doctors had become diggers, trying hard to follow treasure maps in hopes of discovering gold, then announcing—to supervisors, to patients, and, not least, to one another—that we had found it. We sniffed, we poked and pried, we got aggressively "active" (one older doctor's advice: become more active when the patient "resists"), and through it all we had our eyes on the place where the treasure was always located—childhood. If we didn't know, we knew what it was that we wanted to know—*would* know, once we'd made our discovery, once we'd found out the nature of the "nuclear neurotic process" (another phrase in wide circulation).

Dr. Ludwig urged us to let the story itself be our discovery. He even went so far as to joke with me: "Let's see, this is chapter ten we're discussing today." He urged me to be a good listener in the special way a story requires: note the manner of presentation; the development of plot, character; the addition of new dramatic sequences; the emphasis accorded to one figure or another in the recital; and the degree of enthusiasm, of coherence, the narrator gives to his or her account. In the spirit of psychoanalytic self-observation, he urged an even more startling line of inquiry: "There are two more stories, at the very least, that you and I are in a position to construct." He saw a look of mixed perplexity and indifference on my face, and countered with a look of inquiring eagerness—not the silent aloofness that so many of us doctors-in-training had already started regarding as the ideal posture. My silence provoked him to a risk—to guess what was crossing my mind. He apologized: "Perhaps 'construct' bothered you?" "No," I immediately averred, but my voice and lowered eyes gave away the truth. Is that what doctors ought to do—construct stories?

He seemed to hear my unspoken question. He pointed out that many of us are astonishingly willing to embrace all sorts of speculative metapsychology, including its often cumbersome language, yet we bristle at the notion of certain everyday civilities—regard them as in some way a threat to professional achievement. Hadn't one important psychiatrist recently warned us about the dangers of "structuring" an interview, diverting patients from the kind of (verbal) self-presentation they otherwise might make before the doctor? For example, even a preliminary discussion of the weather, we were told, can be "artificial," can "structure" the interview in a direction that might not otherwise have been taken by the patient. True, no doubt: any meeting between two people can end up taking a myriad of forms. Yet one assumes that a certain distress has, finally, compelled someone to take on the role of a patient, and that such personal unhappiness will have a way of making itself known (telling its story), even if thwarted a bit by casual talk. Moreover, we would-be analysts are not beyond perking up here, appearing uninterested there, and, of course, asking questions that set the compass for a given conversation.

Dr. Ludwig was trying to suggest that what had troubled me was not the word "construct" but the word "story," that I associated it with literature, with mere reading—not science, not medicine. He remarked that first-year medical students often obtain textured and subtle autobiographical accounts from patients and offer them to others with enthusiasm and pleasure, whereas fourth-year students or house officers are apt to present cryptic, dryly condensed, and, yes, all too "structured" presentations, full of abbreviations, not to mention medical or psychiatric jargon. No question: the farther one climbs the ladder of medical education, the less time one has for relaxed, storytelling reflection. And patients' health may be jeopardized because of it: patients' true concerns and complaints may be overlooked as the doctor hurries to fashion a diagnosis, a procedural plan. It is not the rare patient who approaches a second doctor with the plea that he or she wasn't heard, that the first physician had his or her mind made up from the start of a consultation and went ahead accordingly with a diagnostic and therapeutic regimen.

In a manner of speaking, Dr. Ludwig added, we physicians bring *our* stories to the consultation room—even as, he pointedly added, the teachers of physicians carry *their* stories into the consulting rooms where "supervisory instruction" takes place. Sometimes our knowledge and our theories (the two are not to be confused with each other!) interfere with or interrupt a patient's momentum; hence the need for caution as we listen and get ready to ask our questions. The same was true for the "case presentations" I was making to my supervisors: I formulated my account of a patient to a particular supervisor in keeping with the way I presumed that doctor was inclined to think with respect to psychological matters. The story I told would be affected by his mind's habits and predilections, *his* story.

Few would deny that we all have stories in us which are a compelling part of our psychological and ideological make-up. What does the presence of those stories imply—for teachers, for doctors, for all who will at some critical point in our lives become patients? That question kept coming to my mind years ago as I began to do fieldwork in the South in the early 1960s, studying the impact of school desegregation on black and white children at the height of the civil rights struggle. Those children were going through an enormous ordeal—mobs, threats, ostracism—and I wanted to know how they managed emotionally. It did not take me long to examine their psychological "defenses." It also did not take me long to see how hard it was for many of those children to spend time with me. They were shy, always clamming up, withdrawing into the protection of their own carefully maintained personal distance. I attributed their reserve to social and racial factors—to the inevitable barriers that would set a white Yankee physician apart from black children and (mostly) working-class white children who lived deep in the segregationist Dixie of the early 1960s. That explanation was not incorrect, but perhaps it was irrelevant. Those Southern children were in trouble, but they were not patients in search of a doctor; rather, their pain was part of a nation's historical crisis, in which they had become combatants. Maybe a talk or two with me might turn out to be beneficial. But the issue for me was not only whether a doctor trained in pediatrics and child psychiatry might help a child going through a great deal of social and racial stress, but what the nature of my attention ought to be. Was I to "treat," or was I to listen carefully, record faithfully, comprehend as fully as possible?

· · ·

Prior to going South I'd spent two years attending young people who had been stricken with polio during the last major epidemic, in the mid-1950s, before the Salk vaccine came into widespread use. I'd noticed how taciturn many of them were; and then, too, I had my (not altogether wrong) explanations. They were paralyzed, were struggling for their lives, were understandably anxious or scared or preoccupied by the daily burdens bad luck had brought to them—so what was the point of talk and more talk?

Not that I wasn't searching out the "defense mechanisms" which had been mobilized and which I was supposed to document. The long stretches of silence, however, the obvious suspicions of me and my purposes, bore down on me. I responded at first by calling on my own "defense mechanisms." *They* were having *their* difficulties, hence the need for patience on my part; yet it was interesting that I hadn't immediately noticed much explicit psychopathology—perhaps its absence was evidence of the patients' use of "denial." Soon enough, I told myself, they'd begin to falter emotionally; that "denial" would break down. Meanwhile, my "protocols" (records of conversations, both taped and untaped) offered their own implicit evidence of denial. Again and again, for instance, I heard remarks about activities in which one or another youth expected to take part. Such a prospect was exceedingly unlikely, I knew, so I called the belief "a function of denial." That awkward phrase was repeated often in the presentations I made at various Boston hospitals. Usually heads nodded in agreement, and a few more abstractions were even pushed on me. Had I thought of the "object cathexes" of these patients—their attachments as they lay in iron lungs or on the beds of the rehabilitation service? Had I reckoned with the superego, considered its role in the "intrapsychic conflict" of these young people? What was happening to the "narcissistic supply" of these victims of polio, to their "body image"?

My replies to such inquiries were, really, mine, not those of my patients. I was making surmises, sometimes with scant proof, only a patient's silence or the use of a word, an image, a metaphor. Might I be clutching at straws, in need of a bit of "help" myself? No one asked, except Dr. Ludwig, who heard my clinical presentations at several grand rounds and at various staff meetings and asked to see me one day. After a few minutes of light banter, he got to his point—that I was all too preoccupied with my theoretical conclusions, meanwhile "missing a great opportunity." He didn't spell out the nature of that opportunity, and I found myself vastly uninterested in having him do so.

I recall trying to change the subject, telling him about a follow-up on a patient of mine whose treatment he'd supervised a year earlier. But he was not to be deterred. He came back to those paralyzed boys and girls I was interviewing, those human beings whose "psychodynamics" I was formulating. "A pity you're not giving us them." What did he mean by that? I knew, but was damned if I was going to say so. I wasn't even going to give him the satisfaction of asking the question. He was *sui generis,* I'd known all along; but as the days and weeks and months had given way to several years of residency training and the beginning of psychoanalytic training, I had begun to realize what a loner he was and how insistent other older psychiatrists were that we should arrive at the correct "conceptualization of the patient's psychosexual status," a phrase I kept hearing and bowing to.

Still, I was fond of that singular doctor, and he was never one to hold back when the matter of professional jargon arose. On the day when he brought the subject up, he gave me a stern lecture on my increasingly opaque way of talking about my patients, and he ended with a plea for "more stories, less theory." He urged that I err on the side of each person's particularity: offer *that* first, and only later offer "a more general statement"—adding as a qualifier, "if you want to, if you need to." I remember flushing. Was he implying that only some neurotic requirement made theory necessary? No, he was not pushing matters that far, he allowed. But most certainly he wanted me to know that a good deal of the time our patients' comments tell their own story, one that can be interpreted by us in ordinary language with no loss of psychological nuance and subtlety. In telling our version of their version of their lives, he claimed, we can make our guesses, indicate our sense of things, without succumbing to overwrought language and overwrought theory.

His polemic did not convince me. Meanwhile, my wife, an English teacher, was doing volunteer hospital work with some of those young polio patients, listening to them speak and finding it impossible to get out of her mind their haunting ruminations, their moral and philosophical questions, as they tried to make sense of their bleak prospects. She was also attending some of the medical conferences to which I'd been invited to present my research. She found it harder and harder to endure in silence what she called the "discrepancy" between "those kids as I know them" and "those kids as you speak about them to your colleagues."

Indeed, once, quite contentiously, she suggested that I wasn't really speaking about them at all; I was, rather, "making reference to them" as I showed "those doctors" how dutifully I could mobilize a theoretical apparatus to the task of "explaining" what was going on. A pause, then: "At times I feel you're explaining *away* those people—and I know you don't want to do so; I know you're really a friend of theirs, and they are friends of yours."

I was jolted and annoyed. I was upset at her reference to friendship. True, we'd all become quite familiar with one another, but they were *patients* of mine, I hastened to remind my wife. Her response boiled down to "So what?" And she added: "All the more reason for you to do justice by them." I didn't like the use of the word "justice." I was working hard; I had done my best to offer my findings to my profession; I had written a paper titled "Neuropsychiatric Aspects of Poliomyelitis," was in the middle of another one, "The Defense Mechanisms in Polio," and was not inclined to feel that my work lacked "justice." An argument took place. I challenged my wife to show me what she had in mind. Two weeks later I was presented with a young woman's story as she had told it to my wife.

The story was deeply affecting, and in its own way poignantly "psychological," full of the girl's memories, her hopes and fears, her constant worries. I was surprised; I felt ashamed; I was envious; I got angry—and I became, to use a word we were wont to use on the psychiatric wards, defensive. (Needless to say, we usually fasten such a word on others.) After I cooled down I went to see Dr. Ludwig with a copy of my wife's chronicle. He was quite taken with her project and wrote to her to tell her so. I was beginning to be convinced, and in time I wrote up more biographical accounts of the lives of some of my polio patients, even letting them tell their own stories through the use of the tape recorder.

• • •

I went to visit William Carlos Williams regularly when I was in medical school, often accompanying him on his home visits. I read anything of his I could find. Often I told him about particular patients, listened to him recall similar medical problems he'd encountered, or heard him describe the troubled individuals he'd come to know—"minds gone all awry," he once said as he thought of a substantial number of men, women, and children he'd met. Like my supervisor, Dr. Williams urged me to tell "doctor stories," a phrase I would use twenty-five years later when pulling together fiction he wrote. Those doctor stories were just that—short fiction meant to evoke the various events, moods, impasses a doctor experiences. I drew in my mind a sharp distinction back then between his somewhat confessional way of writing and the writing I did as a house officer on, say, a psychiatric service of a children's hospital. As a house officer I had to report in detail a given medical or psychiatric reality, and do so in a manner my colleagues could accept and comprehend. I had no time or sanction for embellishments, nor for narrative that didn't draw diagnostic conclusions: what to make of a person's troubles, what to do about them, and with what likelihood of success. Since I had entered a world that had its own language, why not speak it while there? At worst it was a form of shorthand; certainly many were eager and quick to respond to it. Theory was a means of getting to the core of things, focusing precisely on what connects this patient, right here, sitting before me, with others all over the world who belong to a category. I can say, "The patient is phobic"—not a callous or coldhearted or impersonal attitude, but a brief, pointed piece of information shared with another busy professional. Yes; but Dr. Williams had this amplification: "Who's against shorthand? No one I know. Who wants to be shortchanged? No one I know."

He's been dead a quarter of a century, yet I still hear him speaking those words, a poem of sorts to someone having a hard time seeing the obvious. Vintage Williams: the busy, street-smart doctor and the hard-working writer merged into the friendly but tough teacher who wanted his younger listener to treasure not only the explicit but the implicit, all the subtleties and nuances of language as it is used, of language as it is heard. He found moments of liveliness in statements of his patients that were terse, banal, circumscribed. He could pounce on someone's adjective or verb; he could be delighted in the confounding exception that undoes the seemingly foolproof conclusion. Continuities and discontinuities, themes that appeared and disappeared, references, comparisons, similes and metaphors, intimations and suggestions, moods and mysteries, contours of coherence and spells of impenetrability—he spoke of such matters as he brooded over his life as a doctor, a writer: "We have to pay the closest attention to what we say. What patients say tells us what to think about what hurts them; and what we say tells us what is happening to us—what we are thinking, and what may be wrong with us." A pause, then another jab at my murky mind: "Their story, yours, mine—it's what we all carry with us on this trip we take, and we owe it to each other to respect our stories and learn from them."

Such a respect for narrative as everyone's rock-bottom capacity, but also as the universal gift, to be shared with others, seemed altogether fitting. I tried to keep in mind what I heard. I would remember it later as I began meeting children who had to fight their way past mobs to obtain a desegregated education. They, too, knew pain, and had "problems"; they, too, possessed "psychodynamics"; but they were also accumulating stories on their journey, as the New Jersey doctor, close to the end of his, had put it, and the stories were ones that begged to be told.

· 2 ·

Starting Out

THE YOUTH OF FIFTEEN had polio; he would lose the use of both legs. His father had been killed in the Second World War; his mother had died in an automobile accident when he was ten. A grandmother, a sensitive and thoughtful widow in her sixties, had become his main family. The young man had no brothers and sisters. He had two uncles, whom he yearned to see more often, but they were living in Texas and California. (Like him, they were born near Boston.) I came to know this fellow fairly well. I first met him in the emergency ward of a Boston hospital when he came in with a sore throat, feverish, and, alas, weak in the legs. My work with him as a pediatrician gave way eventually to my conversations with him as a child psychiatrist. He was "moody," by his own description, and he was not averse to long talks.

We always started with sports, especially baseball and hockey, his two loves. In time we'd drift toward the hospital scene as he saw it: the nurses, the virtues and faults of various ones, and the doctors, mostly their faults. He regarded us residents with a skeptical eye. We strutted, were all too cocky. "The doctors give so many orders, it goes to their heads." He said those words so many times that I found myself, in retaliation, observing his "hostility." It was hard for him to accept the bad deal life had given him. He was angry, I knew, and needed a target, someone or something to attack, lest he turn all the fury on himself and become depressed. As he described us doctors, scurrying around, always on the move, collaring people with our orders, he seemed wistful. He would look past me, toward a window, and I always hesitated to press our conversation. He seemed gone—his mind was out there, free of his body.

Once, as we talked about that body's prospects, he became philosophical. He wondered whether the soul is always confined to a given body. Might it become migratory? What did I think? I was stupid enough to shun the question and to throw it back at him. He was smart enough to spot my pose—a shrink in action—and irritated enough to give me a dose of his bile. He spoke at considerable length; one remark has stuck with me for the many years since he made it: "If you would tell me what you think, then I could answer better." At the time I wasn't getting any wiser, however. I interpreted that comment as an effort on his part to hide behind me, as it were—to let me know that he would pretend to oblige me by taking cues from me, but not deliver to me what I wanted, his own unvarnished self. He spotted a coy reserve in me at that moment, which must have told him what was crossing my mind. He changed the subject abruptly, instructively. Had I read *The Adventures of Huckleberry Finn?* Yes, I answered, wondering what the question meant. He said no more. It was left to me, during the silence that followed, to figure out what to say, if anything. I waited just long enough to realize that the youth, whose name was Phil, had no intention of proceeding further in any direction. My wife's and supervisors' faces, their voices, rushed to my head. A week or so earlier my wife had urged me to "exchange stories" with the children I was interviewing in the hospital; Dr. Ludwig had agreed: "Why don't you chuck the word 'interview,' call yourself a friend, call your exchanges 'conversations'!"

Suddenly I heard myself talking about Huck and Jim, about the mighty river, about my own experiences as a child when my mother took my brother and me to visit her family in Sioux City,

Iowa, located on the Missouri River. I told Phil that my grandfather used to take me to that river, point south, indicate the destination of the water: the Mississippi, then New Orleans and the ocean. "Those rivers are arteries of the American heartland," he'd tell me—the farmland expanding and contracting, opening up and offering crops, then retreating into the winter lair, and all the while the water flowing, keeping an entire region alive and fertile.

Not brilliant imagery, but enough to shed me of my scrutinizing, wary reticence; enough to involve Phil in a bit of a personal story, which in turn was connected to a reading experience he had recently had, and one I had also had, though about seven or eight years earlier. I was almost ready to tell him how young he was to have read the Mark Twain book—to patronize him foolishly and smugly—when Phil began talking about the book. He had read it as a school assignment before he fell sick. When he'd been in the hospital a week or so and began to realize that he was "really paralyzed" and that his disease might be "for a lifetime," he became morose, more so than others on the same ward with the same disease, for whom the bad news had yet to sink in. All he could think of was "the black space" of his future life. But a teacher came to visit him on a Saturday afternoon, and the result was a reacquaintance with the Twain classic. Not that young Phil relished the idea at first. Here is how he described what happened (his remarks have been edited and on occasion reconstructed because my tape recorder intermittently broke down):

"I was surprised to see him [the teacher]. I'd liked him, but he was gone from my life, the way a teacher is when you go on to the next year of school. I guess he heard I was sick. We all knew he was a softie! Some of those teachers don't give a damn for you as a person. They talk to the back wall, and if you hear, fine, and if you don't, you flunk! This guy we all knew—he was different. I guess I didn't learn how different until he showed up here.

"He came in and smiled and said hello. I was surprised. I said hello back. I didn't have anything more to say, though. He was quiet, too. I was glad! I was tired of people coming and expecting me to talk with them. I wanted to lie there and think. I felt like crying, but I didn't; I couldn't; I think I was afraid that if I started—once I started, I'd never stop. He just sat there and smiled; then he asked if he could go get me something—food, or a glass of juice. I was thirsty, and I said, 'Yes, orange juice'; and he left, and came back with orange juice and with some peanuts. I liked that, the peanuts. I used to nibble on them a lot before I got sick. I remember my mother saying they were better for me than chocolate. I got a little choked up then, thinking of her and the peanuts and looking at my legs. No more baseball. No more hockey. No more walking, either.

"I saw him looking at the magazines I had on the table near my bed. He leafed through them; then he asked me if he could bring me some books, maybe. I shook my head. I didn't want any books. I was beginning to think I didn't want any teachers here either —*him.* Then he said he was going to go! I guess he'd read me! I felt like I was going to cry, but I didn't know why. I was afraid of breaking down in front of him. I tried to tough it out. I became flip. I joked about having a ball when I came back to school—speeding down the corridors at sixty in a wheelchair. He smiled, but he didn't laugh as much as I did. I knew when I was laughing that it was fake. In a minute he was gone—and then I did cry. I didn't even want to see another day. It was raining outside, and I was crying, and my legs were useless, and I haven't even graduated from high school, that's how young I am, and all I can see ahead is those rehabilitation people, and nurses, and my grandmother looking so worried, and she looked so sick, once I got sick. For a while I thought she was going to die, and then there'd be no one.

"He came back a few days later; he had this book under his arm. He didn't push it on me. He stood there and talked, small talk, and I talked. After a few minutes there was nothing more to say. Suddenly, without saying anything, he up and left. I thought it was strange, the way he left. But he hadn't left; I mean, he came back. He had orange juice in one hand, a glass, and peanuts in the other. I couldn't help smiling. That was the first smile on my face, I think, since he'd been to see me. We talked a few minutes more, about the lousy weather, and then he said he was going. He shook my hand, and just as he was saying good-bye, he took the book from under his left arm with his right hand and put it on my table. He didn't say anything, and he was out of the room before I could say anything.

"I was really curious to see which book he'd brought. I looked, and saw it was the Mark Twain one, *Adventures of Huckleberry Finn.* I started flipping through the pages. I wondered why he brought it. I'd already read it—in his class, last year. What was the point? I guess I was a little annoyed with him. I wondered what was wrong with him, at first. Why that book? What's he got in mind? I asked myself those kinds of questions. I didn't go near the book for a few days. It was just there, with the magazines my grandma brought. I didn't read them much either. I'd look at the pictures, and I'd read a paragraph—and you know what? I'd get sick to my stomach. I'd feel like puking. I thought it was part of the sickness. I told the nurse, and she told the doctor, and he asked me, and I explained to him what was going on. He examined me, and told me it was all in my head. I joked with him: I said, 'All'?

"When the doctor left the room, I decided to pick up that book; so I did. I flipped through the pages, and then I started reading it, and then I didn't want to stop. I read and I read, and I finished the whole book that night; it was midnight, maybe. The nurse kept coming in to tell me I should put my light off and go to sleep because I needed my rest. What a joke! Are you kidding! I said to her. I'm going nowhere. I'll be in bed for the rest of my life. What difference does it make to me, night and day? She backed off. I read, and when I was done with the story, I felt different. It's hard to say what I mean. [*What do you think happened?*] I can't tell you, I can't explain what happened; I know that my mind changed after I read *Huckleberry Finn.* I couldn't get my mind off the book. I forgot about myself—no, I didn't, actually. I joined up with Huck and Jim; we became a trio. They were very nice to me. I explored the Mississippi with them on the boats and on the land. I had some good talks with them. I dreamed about them. I'd wake up, and I'd know I'd just been out west, on the Mississippi. I talked with those guys, and they straightened me out!"

At that point he paused for a long time. He shook his head. He stared out the window. Then he abruptly put a question to me: "Have you ever read a book that really made a difference to you—a book you couldn't get out of your mind, and you didn't want to [get out of your mind]?"

Yes, I said, and knowing he wanted an example, I told him: *Paterson,* William Carlos Williams' long poem. We got into a long talk about Dr. Williams' medical work with mostly poor and working-class people, about his effort through stories and poems to understand America's social history and moral values. He asked for examples, which of course I didn't have on hand. But he was obviously setting the stage for another conversation. I got my Williams books out of a box, brought a couple of them to his room the next day, and read from the first two books of *Paterson* and from various poems Williams had published in the course of his long writing life. I will never forget the direction of our discussion afterward. Phil wondered whether Williams would ever have been able to accomplish what he did, were he not inspired by what he saw all the time as a practicing physician. Then he wondered whether Mark Twain, whose life he had briefly studied, would have been able to do the kind of writing *he* did, had he not been such an inveterate wanderer before he found himself having much to say. The reason for Phil's interest in pointing out the connection between art and life was not too hard for me to comprehend—or for him, either.

He began musing out loud about his future prospects, with discouragement and dismay. In reply, I pointed out that writers are constantly creating their own worlds, not necessarily needing to travel far and wide in order to gather the particulars for so doing. He once more wanted examples, and our next minutes were taken up with Jane Austen and *Pride and Prejudice,* which I'd read in high school and which his closest friend, a year older, was about to read at the behest of an English teacher. Well, Jane

Austen was a novelist, a writer—lucky to be able to achieve what she did, living the life she did. Things would be different for him. He was no writer, would be no writer, had never even thought of becoming one. Now, significantly paralyzed, he could not even be the day-to-day athlete he'd been; nor did books seem the most inviting of alternatives. He politely but firmly reminded me that he was not "the greatest of students," that he was a "slow reader," that he was struggling with his own worries and terrors, not those described by a novelist in a story: "I wish one of those writers had written about the mess I'm in!"

I did not, then, try to come up with a novel that might pass his muster. Even if I had known of a novel with a polio victim of his age, sex, and background as the hero, I would not have mentioned it at the time. His complaint went deeper; like Job, he was puzzled in the most profound way possible and wanted to find his own voice, use it to make his own plea, his own cry, though he had already begun to regard the world as largely indifferent to him and his situation. I decided to await his decision: whether to do some reading as a means of reflecting indirectly, but with emotional resonance, on his personal situation; or to reject such a way of trying to come to terms with his ongoing situation. A week later, as I was talking with him in his room, I noticed a new box of candy on his bedside table, and underneath the candy box a book. The title was not immediately obvious; I had to move toward the window on a pretext—a bit of sun in my eyes, so best to pull the shades. As I did so, I saw that the book was *The Catcher in the Rye.* I didn't say anything; neither did Phil.

A few days later, as we talked about the rehabilitation efforts taking place, Phil suddenly changed the subject: "I've discovered a book that has a kid in it like Huckleberry Finn." I said nothing but looked interested. He asked me, "Have you ever read *The Catcher in the Rye*?" Yes, I answered. "Do you see what I mean about Huck Finn and Holden Caulfield?" Yes, I answered. Silence. I got alarmed. Why wasn't I feeding our conversation? Why my terse yes, two times spoken? But he began a lively monologue on that novel, on Holden, on Pencey Prep, on "phonies," on what it means to be honest and decent in a world full of "phoniness." Holden's voice (Salinger's) had become Phil's; and uncannily, Holden's dreams of escape, of rescue (to save not only himself but others), became Phil's. The novel had, as he put it, "got" to him: lent itself to his purposes as one who was "flat out"; and as one who was wondering what in life he might "try to catch." He lived on a city street rather than near a field of rye. He was not as utopian, anyway, as Holden. But this youth had been removed by dint of circumstances from the "regular road" (his expression) and he was trying hard to imagine where to go, how to get there. *The Catcher in the Rye* enabled him to return at least to the idea of school—to consider what kind of education he wanted, given his special difficulties.

He had been getting some tutoring in the rehabilitation unit of the hospital. Now he began teaching himself—leaving the building for Huck and Jim on the Mississippi, for repeated excursions to meet Holden. A friend of his invited him to expand his travels, to visit Ralph and Piggy and Jack and Simon on the tropical island in William Golding's *Lord of the Flies.* But Phil resisted that invitation; the book, brought to him by the friend, remained unread. He had glanced at it, seen its charged symbolism, its mix of hard realism and surrealism. Huck and Holden stirred him, brought him to reflection; Ralph and the band of boys on the island were "not for me." When Phil said that, he looked at me and saw my curiosity rising; he decided to give me a terse explication, one I would never forget: "I'd like to leave this hospital, and find a friend or two, and a place where we could be happy, but I don't want to leave the whole world I know."

My wife was quite taken by Phil's way of putting those books into a perspective that suited him. He was calling on certain novels in his own manner and turning away from others for his own reasons. (Phil also rejected the detective stories his friends brought to him, and the Westerns.) A week or so later I heard him again talk about Huck and Jim and Holden. They had become, for him, a threesome. Rather, he had joined them; they were a foursome. His misfortune had evoked in him a wry, sardonic side. He was quick to notice hypocrisy or deceit in the world as it came to him—on television, in the newspapers, in reports from friends and family members, and through hospital personnel. One particular doctor especially offended him, reminding him of a certain teacher and also of an uncle, his mother's older brother. They all "pretended to be nice," but were (in his judgment) "phonies." How he loved that word; what palpable pleasure it gave him! Once, using it, he must have noticed something cross my face—an expression in my eyes, a tightening of my face—and he must have guessed that I thought his use of the word was significant and perhaps inappropriate. He called me on my heightened response to the word "phony"; he told me that both Twain and Salinger were warning the reader to take a hard, close look at the world. If I did so; if I read those two books as he had recently done; if I would "stop and think," then a recognition would descend upon me, too—or so he hoped. He loved the blunt, earthy talk of Twain, and Salinger's shrewd way of puncturing various balloons. He didn't like being paralyzed; but he did like an emerging angle of vision in himself, and he was eager to tell me about it, to explain its paradoxical relationship to his misfortune: "I've seen a lot, lying here. I think I know more about people, including me, myself—all because I got sick and can't walk. It's hard to figure out, how polio can be a good thing. It's not, but I like those books, and I keep reading them, parts of them, over and over."

· · ·

Several years later, when my wife and I were talking with the black and white high school students in Atlanta who were initiating school desegregation—she was by then teaching in a Georgia high school—we had conversations with a number of young people about their participation in a major historical crisis. We were not loath initially to ask direct psychological questions, but we soon learned how reluctant many of our informants were to reply. True, the briefest of responses could always be grabbed by me and turned over (and over and over) until a matter of major importance might be interpreted as lurking behind a spoken moment of indirection, which is the way it often goes in psychoanalytic psychiatry. Yet I yearned for longer, more candid talks.

At one point, when discussions had flagged for several weeks, my wife suggested I be more patient with the students, not press them so hard for a psychological declaration. Two days later, as I sat with a seventeen-year-old white high school senior girl who had been made uncomfortable by the presence in the classroom of one black girl, I found myself hearing about a "tough" English teacher who had assigned three essays, one after another, to her students, among them Laura, the speaker. What did I think of a teacher who seemed so relentless in her demands, so merciless in her grading? Why, the highest grade she so far had given was a B–! Moreover, she kept telling the class this: "You're not putting enough of yourselves into the classroom discussions." My wife asked Laura what she was then reading for that class. The answer: "*Pride and Prejudice*—a really hard book to follow." My wife asked a few more questions. The gist of Laura's complaint was that "there's a lot of talk in the novel" but that it didn't seem to be "going anywhere." She wanted more "action"—and, really, a more contemporary plot.

We eventually had a long and rather intense exchange in which Jane Austen's intentions were examined as thoroughly as each of the three of us could manage. When Jane Austen wrote *Pride and Prejudice* she could not have known its possible value for Americans caught up in racial conflict. The novel was published in 1813; the three of us were discussing the novel exactly one hundred fifty years later, in 1963. My wife pointed that out to Laura, to her noticeable indifference. She, like Phil, had read Salinger's *The Catcher in the Rye* (at the time Salinger was a literary hero for many young people, and at the height of his writing career) and had found it "great." She pitted Salinger against Austen, talked about how much more the former "saw" than the latter. My wife skillfully asked Laura what Salinger saw, and was told, in sum, that he penetrated the surfaces of human behavior, saw the conceits and deceits to which we are all heir. In her words: "The world has lots of phonies, and when you finish *Catcher in the Rye*, you can spot them better."

Had that achievement been an immediate bonus? Yes, Laura averred. She hardly had finished the novel when she was able to spot "the frauds" among her classmates and teachers. Even her relatives fell under a new and stringent scrutiny: "I have this uncle, and he's a big phony. He thinks he's better than anyone else, just because he's made a little extra money. He bought a house out in Buckhead, and he doesn't want to recognize us. He had us all out one Sunday afternoon, the whole family, and he was taking us from room to room and putting on the dog. I was disappointed in my daddy [the uncle's brother] because he fell for the act, hook, line, and sinker. Daddy should have known better; but he's a nice guy. That afternoon he was forever staring and oohing and aahing, and he near turned green with envy. On the way home Daddy became an amateur philosopher; he said God chooses some people to be rich, and that's how it is, and you have to settle for your luck, and ours isn't all that good, so that's too bad, but if you just smile and keep going, then you'll be fine; it's when you eat your heart out that you can get in trouble. I felt sorry for him. Actually, I was mad at him. He was being foolish. We never even should have gone there, to Buckhead. It's way above us, the place, and when you see it, you feel real low, and it's stupid to let that happen. Up there, people don't respect you for your honest-to-God self; they're *measuring* you—where you live, and what you're wearing, and your accent, all like that. Daddy used to tell us when we were small kids that the worst people are snobs, and there he was paying court to one, his own brother. No wonder my mother got a splitting headache, and even three aspirins wouldn't stop it. She went to church and prayed to Jesus, please, to heal her head, and He did."

We talked about money and social status, about what happens to people when they climb higher and end up living in places like Buckhead. We talked about honesty and phoniness. We talked, too, at my wife's quiet but firm insistence, about *Pride and Prejudice.* Laura was not unable to comprehend the prejudice in the novel—the class-conscious, condescending side of Fitzwilliam Darcy, not to mention that of his snotty, self-important aunt, Lady Catherine de Bourgh. Moreover, as we conversed, Laura pointed out how "intimidated" the victims of "pride and prejudice" can become. She called upon personal experience again: "I've seen my mom and my daddy become different people when they're with that uncle of mine. They're scared of him, I think. They try to get him to be friendly to them—and it's all so phony. He's got the 'pride and prejudice.' He tries to forget where he came from. But my folks, they're trying to get into his boat—I mean, his big Lincoln Continental. It's sad the way people get lost when they run up against other people. I saw a picture of Daddy and his brother when they were little boys, in first grade and third grade, and they had their arms around each other, and they had big smiles on their faces, and they were—true brothers. Now they're strangers. So you see, it doesn't always pay to grow up!"

Such a viewpoint was obviously nourished by Salinger, with his profound distrust of so much that gets called "the adult world": a fiercely competitive, materialistic culture. Salinger's jabs at the crassness of corporation lawyers and businessmen had been virtually memorized by her. She paraphrased one of Holden's remarks for us: "I don't know how much Dad makes, but it's a lot. He's a corporation lawyer, and those guys really haul the cash in." Another line of Holden's: "What you do is study so you can learn enough to be clever enough to buy a goddam Cadillac one of these days, and you've got to pretend you really care that the football team loses, and you talk all day about gifts and liquor and sex, and you stick together in those dirty goddam cliques."

I was stunned at how earnestly she'd tried to commit such passages to memory. True, she wasn't by any means word-perfect. But she knew what she was trying hard to remember, and why. She had made no such effort with *Pride and Prejudice,* though she certainly did remember the gist of the famous opening lines. But as Laura and my wife explored that novel, its assault on phoniness became more evident to Laura. "Even back then," she observed, "the same hypocrites were around." Eventually the discussion shifted to the topic of school desegregation—the hurdles faced daily by two black students, and the "confusion" she and other white students felt. In this regard she was much better able to use *Pride and Prejudice* than *The Catcher in the Rye.* She commented on the differences among the Bennet sisters in the Austen novel—their distinctive personalities, temperaments. How did I (an "expert") explain such a development? I reminded her of what she'd been telling us about her father and her uncle; and she was surprised at the obvious parallel. She was able to announce, at one point, that *Pride and Prejudice* in fact offered parallels of sorts to the circumstances in her high school: "I guess lots of us [white] folks are as proud as can be, and we've got our prejudices. I'll have to admit it. But if you think about it, you mustn't say it's all good on one side and all bad on the other. The Bennets are poor compared to Mr. Darcy and his aunt, and you're really on Elizabeth Bennet's side, but you're not on Mary Bennet's side, or Lydia Bennet's side. If you ask me, you're not on their mother's side, either. And Mr. Darcy is a very good person in a lot of ways. I like him better than Mr. Bennet or George Wickham. Darcy has to work on Elizabeth to see *him* and not some rich aristocrat. There's prejudice going in both directions; and there's pride going in both directions. Now that we talk, I remember one of the colored kids; she was walking near me, and I said hello, and she looked straight ahead. I think she was afraid I was cussing her out. I wasn't, but I didn't speak loud enough, so she probably didn't hear me. I spoke real soft because I was afraid to be friendly to her. It was a bad scene all around! But she was as locked in as I was. I raised my voice and said hello again. She still ignored me. She thought I was setting her up, I know. I wanted to grab her and say: Look, sister, no use you being a snob, just like the rest of us! I've seen her with the other one [a black young man], and she is pretty tough with him. [I asked her what she meant.] I mean she won't give him the time of day. She snubs him the way us whites snub her! I heard her parents are better off. Her father works for the post office. His father is a janitor, and he drinks a lot."

Laura told us how she had learned such information—through a teacher, who was all too gossipy. She also told us how carefully she and her friends had watched those two black students and watched one another with them: who said what, if anything, to them in class, while walking down the school's corridors, in the lunchroom. My wife suggested that she, that they, were doing as Jane Austen did—telling stories to one another in which plots are developed, characters portrayed and plumbed, incidents recounted. Moreover, Laura had criticisms to make of one classmate's rude behavior to the black young man. She tried to explore the reasons for another classmate's effort to talk with that same young man; the boldness and yet the subtlety of her account would not have been unworthy of Jane Austen: "She's wrapping us all around her finger! She knows that for a white girl to be friendly with a colored kid, a boy—that's going to get everyone's attention, you bet. She's an actress! She loves people watching her. She's as cool as can be: she doesn't push things; she doesn't get real palsy with him, but she says her 'hello,' and she says her 'hey, how you doin'?' He's made nervous, I can tell. He tries to avoid looking at her, so that she won't talk to him. But she positions herself so he's got no choice, and then you can hear a pin drop. She doesn't lose it [lose her cool], though; she's quick to pick up talking with one of us, and to act as if nothing happened, nothing. I heard a teacher tell another teacher—he'd seen her giving him [the black student] a 'hi'—that 'the girl belongs in the theater.'

"I guess it's all right, though; I guess it's good, actually. You can't always judge people by their motives. She's helping to break the ice, and that's more than I'm doing! I'm too scared. I'm too shy. I wonder which one of those Bennet sisters I resemble. Jane, maybe. I'm not as proud as Elizabeth. She was as proud as Darcy, and as prejudiced. I hope I'm not one of those selfish Bennet people, or like that aunt of Darcy's. When you read a novel, even if it's one you don't like so much—or you have a hard time getting into it, and staying into it—you can't help trying to figure out which character resembles you; or I guess it should be the other way around: which character do you resemble most? When I read *Catcher in the Rye* I kept thinking of myself as Phoebe Caulfield, even though I knew I wasn't smart like her. It's just that I loved Holden, and he loved her in his way (he truly respected her); and I wanted his respect. I get lost in those books, you can see! It takes me an hour sometimes to get my feet on the ground after I've been reading a book like *Catcher in the Rye.* Even *Pride and Prejudice* can take you over."

One day Laura talked with her teacher about *Pride and Prejudice,* about Elizabeth and Darcy, about the "racial problem" which dominated everyone's thinking that academic year. The teacher reminded Laura that she and her fellow students were not the only ones being asked to stop and think about old customs and privileges in a new way: "She told me that she was fifty years old, and she never expected to see this desegregation happen, and she was trying to do the best she could. She told me she loved teaching, and she hoped she could get us all to 'use' this desegregation so that we learn from it. I wasn't sure what she meant, but later in the day I could see she had been talking to herself as much as to me, because I realized how nervous that black boy in our English class made her. She kept staring at him. It made me nervous, seeing her made nervous.

"I think last week we had a breakthrough. She came into class Monday morning as if she'd had a long vacation, and she was just raring to go. I could tell by her movements—she was darting here and there—that she was back to her old self. She handed out this story, and told us it was important, and we should read it carefully once, then a second time, and we'd discuss it in class. None of us had ever heard of the author. But there was something about the way she [the teacher] talked about that story—talked it up—that made me feel it wasn't just another story; it was some important message for us in the class. So, I took it home and read it twice, and tried hard to figure out what the author was trying to say. That's what she [the teacher] has always said to us: 'When you read a story, enjoy it, but you should also try to understand what the author is trying to tell you.' "

· · ·

From Laura and from her English teacher as well, I heard a great deal about the impact of that story, Lionel Trilling's "Of This Time, of That Place." At the beginning of the story Joseph Howe, a teacher at Dwight College and a writer, is trying to establish himself in the world, gain attention and approval from an audience of readers and from his colleagues. The author observes: "The Bradbys would be pleased if they happened to see him [Howe] invading their lawn, and the knowledge of this made him even more comfortable." A man is struggling for self-respect and self-knowledge but links such achievements to the way others regard him. When he assigns to his students an essay titled "Who I Am, and Why I Came to Dwight College," he is indicating the nature of his own moral inquiry. Mr. Howe tries earnestly to figure out how to live his life.

Laura's teacher boldly identified herself with the teacher in Trilling's story. She reminded the class that teachers also have their difficulties, their challenges, especially at moments of public turmoil. She pointed out that students can help teachers intellectually and personally, not to mention that students can help one another. In Trilling's story Howe has significant encounters with two students, one named Tertan, the other named Blackburn. At one point Tertan states, "Tertan I am, but what is Tertan?" Unsure of himself, the boy rambles in a disjointed fashion. His jumbled words tell what takes place in his confused mind. He is referred to as "mad"—yet he can be highly perceptive, and he intrigues his teacher and some of his fellow students. On the other hand, Blackburn is aggressive, manipulative. Tertan lacks control over himself. Blackburn calculates with evident precision the lengths to which he can go: he dares to blackmail Howe for a good grade—threatens to lie, place him in public jeopardy. Howe stands up to him, then backs down. Blackburn, too, is "mad," determined at all costs to succeed. As the teacher takes the measure of those two students, he begins to come to terms with two aspects of himself. Like Tertan, Howe is not able to command his words, his manner of self-presentation. Like Blackburn, Howe wants to rise up, make a mark in the world.

It was not hard for Laura's teacher to prompt members of the class to stop and consider moral issues. Soon enough the class was discussing itself, and in doing so, thinking about the capacity each of us has to intimidate, to be intimidated, to be driven half crazy by this life's perplexing and sometimes irrational nature, and to drive others toward madness, meaning toward anger as well as irrationality. The teacher went further, pointing out that "black" has many connotations, as does "white," and that often a perceived sense of another person (pro or con) tells a good deal not only about our vision but about us, the beholder. The class responded, hesitantly at first, but with increasing commitment of both head and heart. In the teacher's words, as offered to my wife, who was interviewing a number of her fellow teachers, all caught up in the city's school desegregation crisis: "I thought for a while the Trilling story would be a dud. It has its precious side—when you think of this public high school, and the kids who come here, from ordinary working families. But the beauty of a good story is its openness—the way you or I or anyone reading it can take it in, and use it for ourselves. For a few minutes, I must confess, I thought I'd really made a mistake. The class was silent. I began to think they didn't like the story. But then a boy raised his hand and asked what Howe's problem was—and we were off and running. By the end of the time we had, hands were still flying in the air, so I suggested we devote the next day to the story, and the day after that, too, if necessary.

"I remember being told in college by my favorite English professor that there are many interpretations to a good story, and it isn't a question of which one is right or wrong but of what you do with what you've read. I told my class that here we are, in our time and our place; and I asked them, please, to stop and question themselves, the way those two students helped Mr. Howe to question himself. I asked them to look around—and look not only at 'others,' but at what they were doing with those 'others' in their minds. Oh, I wasn't giving a lecture in psychology! I noticed that when we gave those 'psychology of prejudice' lectures at the beginning of the school year [part of a social studies curriculum geared to race relations], lots of students were bored or indifferent. They fell asleep—I mean, the words fell on deaf ears, the way they so often do during lectures. But when we got going on that Trilling story, there wasn't a heavy eyelid in the class! Maybe it was me, putting a lot of myself into the class."

Trilling's title for the story is obviously ironic. True, one teacher is struggling at one time, in one place. But Howe's struggle is shared by millions of us who want to know where we belong and with whom we will spend our lives. Like the students he teaches, Howe is himself ready to graduate, to become his own man, to join the "procession" on its way to the challenges life will offer after school ends. He wants to get along with the world in a reasonably sane fashion without losing the special and odd vision of the artist. Tertan possesses some of that vision; hence Howe's feeling of affinity with him. Tertan's very name—the third or odd way—is suggestive. The students were able to use Tertan as a means of discussing not only Trilling's story but their own. In Laura's words: "We went round and round the subject of the majority and the minority, the conventional and the unusual. After a while, I think some of us began to realize that the racial situation in our class could be connected to the story; I mean, we all had been feeling strange about this desegregation we were going through, and all eyes were on those two black kids; talk about being out of place and odd! I know there have been times when I've felt a bond with them—when I realized how odd I myself feel, and how lonely: there are days when you feel no one understands you, and when you don't care about anything, or anyone, you just want to get by, and you pay attention to no one. That's how I think those colored kids feel, and if you haven't been there yourself, even if you're white, then you're pretty lucky. And sometimes, it's those 'down' moments when you do your best thinking. You're most yourself. I think Trilling was getting at some of that in his story."

· · ·

My wife delighted in the response those young people gave to another story never intended, certainly, to assist schools struggling to become desegregated. In her own teaching she was using Tillie Olsen's new book *Tell Me a Riddle* to great effect. She had urged it on Laura and others she knew outside her classroom. The four stories in the book all addressed the experience of loneliness. The racial tension in the schools, as we had begun to realize in the early 1960s, caused many students, white as well as black, to feel not only perplexed or fearful but lonely. As Laura pointed out to us: "If you're white, you can't just forget this trouble. It affects everyone, and it makes you wonder what you believe, and what you should do. We're all out there on our own, in a way, one of the teachers said, and she was right."

When the Atlanta students of both races read Olsen's stories—and when their teachers did, too, for the first time—the reaction was invariably strong and almost always favorable. "I Stand Here Ironing," placed first in the collection, stirred much emotion in those Atlanta high schoolers at a time when there was no formal women's liberation movement and when Olsen herself was hardly known. The woman who is ironing is poor and hardworking; her children have not had an easy time of it. The story, which tells of her thoughts as she does a household task, takes the form of an intensely personal and searching reply to the challenge of a social worker who called her on the phone and said: "I wish you would manage the time to come in and talk with me about your daughter. I'm sure you can help me understand her. She's a youngster who needs help and whom I'm deeply interested in helping." What follows is a mother's self-scrutiny as she recalls the past, casts a candid look at the present, and anticipates the future.

The narrative is alive with a woman's frustration, sadness, and bitterness—but also her persistence, her toughness, her determination to survive. For those Atlanta students, and for other high schoolers we have met since then, the story has special meaning. The daughter in "I Stand Here Ironing" is herself of high school age, a mix of talent and long-standing "problems." Olsen affirms the girl's complexity, the ambiguous nature of her personality: she has a dour and melancholy side, but also possesses notable talent as a mimic. The story prompts young readers to look at their own past—to take stock of the troubles they have had, and the opportunities, too, and to reflect on how they have managed in the less than two decades of their lives. My wife's students, and the young people we both were meeting in Atlanta, welcomed that story as a godsend—a stimulus to "balancing the books," as one eleventh-grade black student put it to us. He assumed that the characters were white, and made a point of letting us know that "never before" had he so identified with people in a "white family." Moreover, the daughter, Emily, as he reminded us, is "a white girl," and for a while this sturdy and athletic young man worried on that score: "I thought to myself: what are you doing, getting yourself into her head? But I couldn't stop it from happening! I'm not the oldest, like she was, but I put in a lot of time with a church nursery, because my mother was working all day; and then I got sick, real sick, and they thought I'd not make it. I can remember my mother looking at me and saying that if I didn't put on some weight, fast, I'd be 'no more for this world.' I asked her, 'Where am I going, Momma?' She started crying, and so did I. That's what happened to me when I read the story for the first time: I started crying, and I didn't know why. Strange!"

Such moments of resonance across racial and sexual lines kept arising as my wife went through the essays she received from her students and listened to the remarks her teaching colleagues made as they, also, embraced Olsen. Olsen herself—a working-class woman whose desire to write had to be set aside for work and more work—caught the interest of students and teachers alike. Many of the high school students my wife was then teaching were quick to take the side of the much-harassed mother in the story, who was trying to make do under difficult circumstances, and of her child, whose "troubles" struck those readers as only normal. They found it odd, actually, that "anyone" (the anonymous school guidance counselor or clinic therapist) would take notice, would ask for a chance to have a conversation. Usually, as a matter of fact, the students were critical of that therapist. The rhetorical questions kept coming up: "How does someone in a clinic know how you live, if you're from another world, and they never come near where you live, and they don't try to find out, not really?" A black girl's essay offered the corroborating experience *her* family had—years of contact with city officials who put her paralyzed mother through many a demeaning inquiry.

The Olsen story "Hey Sailor, What Ship?"—also in the collection *Tell Me a Riddle*—was even more strongly appreciated by those Atlanta students, even though its narration may make more demands on the reader. My wife was surprised, in her early years of using the story, at the emotional force of the students' response. I think we both had failed to consider how many people struggle with alcoholism and how significantly that struggle bears down on children. But the story is not primarily psychological; it poses tough and serious moral questions. Whitey's alcoholism has in many ways hurt him, yet he is an altogether decent and generous person. He sees the world through the blurred, tipsy daze of a long life wedded to bottles of various shapes and contents. But he digs deep in his pocket for cash and finds it not only for bartenders but a host of friends. There is an innocent vulnerability and hilarity to him; and he wants to share his warmth and spontaneity with others, especially his old friends Lennie and Helen and their children. They are his family, but they gradually become something else to him and to the reader. Jeannie, the oldest daughter, worries about the example he sets for her younger siblings. The mother responds poignantly, unforgettably: "They don't hear the words, they hear what's behind them. There are worse words than cuss words, there are words that hurt. When Whitey talks like that, it's everyday words; the men he lives with talk like that, that's all."

But Jeannie will have no part of such an explanation. She deplores his vulgarity, this man who has loved her, showered her with gifts, played with her, livened up her family's life for many years. Even though her mother reminds her of the vulgar language she and her friends use, the girl scorns him while at the same time wanting his money: "Why didn't Daddy let me keep the ten dollars? It would mean a lot to me, and it doesn't mean anything to him." No, the mother won't accommodate such a presumptuous self-centeredness, a thinly rationalized greed, and in consequence the daughter tunes up her scorn, calls him "just a Howard Street wino." Try as the mother does to appeal for understanding, if not compassion, Jeannie cannot be budged. "He doesn't belong here," she insists. The mother points up another irony—Jeannie can find another kind of alcoholic acceptable: "You think Mr. Norris is a tragedy, you feel sorry for him because he talks intelligent and lives in a nice house and has quiet drunks."

So it goes—a child growing up, learning whom to respect and why, whom to deny all respect and why. The story is not only about an alcoholic, but about the rest of us, who have other problems. At the end Whitey walks the port city's pavements, headed back to the ocean that has been his lifelong home: "He passes no one in the streets. They are inside, each in his slab of house, watching the flickering light of television. The sullen fog is on his face, but by the time he has walked to the third hill, it has lifted so he can see the city below him, wave after wave, and there at the crest, the tiny house he has left, its eyes unshaded. After a while they blur with the myriad others that stare at him so blindly."

With those words Olsen invites us to comprehend a narrowness of vision that is not induced by alcohol. Whitey joins hands with the nameless "invisible man," Ralph Ellison's protagonist, and Dostoevsky's "underground man"—outsiders whose narrated lives tell us a good deal about our moral values and assumptions. For youths of fifteen or sixteen Whitey's story is accessible and affecting. We found its moral energy especially contagious among our white and black high schoolers, who had greeted direct discussion of race with suspicion, if not irritation or bore-

From the desk of:

7/8/98

RE/MAX

Transcription

Carol 11.0 hrs

MH 8.0 hrs

19.0 X $15.00 = $285.00

Thanks

Diane

RE/MAX Above the Crowd!®

An Independent Member Broker

dom. Similarly, in the North, in recent years, we have found that many young people get "caught" by Olsen's story, much to their own surprise and even their teachers'. A white high school student of sixteen in a suburban Boston school remembered his involvement with the story in the spring of 1983: "I found it hard to read her [Olsen] at first. The teacher had to explain a lot, the way she writes. But once you catch on—well, you get caught. She has this way of getting you to stop and think, and the story is so packed, you can't put it out of your head so easily. I'll read a novel, and I don't remember it the way I remember those four stories, especially 'Hey Sailor, What Ship?' I suppose a lot has to do with my father, his drinking. [*Is he an alcoholic?*] No. But he really goes under, starting at six o'clock, when he gets home. Mum tries to feed us separately, so she can take him on and try to keep him from 'blowing.' [I asked what he meant.] That means his temper exploding. He'll throw stuff around and curse his boss. He'll curse anyone who gets in his way or comes to his mind! We hide and listen.

"When I read that story I thought of my dad—like that Mr. Norris, though Dad is a lot noisier, I'm sure. But it's not just that Dad drinks a lot. The way the teacher taught it [the story], you stopped and asked yourself a hell of a lot of questions. You begin to realize that we're all in trouble, one way or the other. The black kids [bused to his high school from a Boston ghetto] are obviously in trouble. My dad is in trouble, but no one on the outside knows. We're living inside our 'slab of house.' I'll never forget those words, and the 'flickering light of television.' My sister and I—we hole up; we go upstairs and try to fight his noise with the TV. We turn it up and up. Once he came running upstairs and screamed at us that we were making all this noise. *Us!* I lost my cool. I figured, let him come after me; let him kill me; I don't give a god-damn. I gave it to him; I told him; I said: 'You're telling *us* to stop the noise! We hear your noise all the time. We're trying to stop hearing it. Why don't you tell us what's bothering you? Why don't you find out, and tell yourself? Then you'll stop making your noise, and we'll all be quiet.' That's what I said. I'll remember every single word for the rest of my life, until I die, I'm absolutely positive. And I'll remember him, standing there. He had his glass in his right hand, and it was drained, only the wet ice—a little orange color to it, the trace of the Scotch. He just looked at me. The words got through the booze barrier; he heard them. He didn't say one word. He looked me up and down, I guess deciding for the first time that I wasn't a little kid anymore. Then he turned around and went downstairs, and he didn't say another word that night. I was shaking afterwards; I felt like crying, but I didn't; I couldn't."

For all the high drama of such an occasion—with the implications of a possible psychological breakthrough—that home's turmoil persisted. Still, even though the father's alcoholic bluster continued, the son's life had clearly taken a turn. He tried to get both his parents to read the four stories in *Tell Me a Riddle.* Neither would do so. Not that he was surprised. Actually, what surprised him was his own devotion to "Hey Sailor, What Ship?" and its continuing capacity over a span of months to affect him. He was, he announced to a teacher one day, no "convert to literature"; he was headed for a "career with computer technology." But an intense classroom experience, as it connected to his everyday home life, had most definitely touched him, prompted him to take a look around.

The other Tillie Olsen stories in that quartet ("O Yes" and the title story, "Tell Me a Riddle") only heightened the impact of the book on his mind and heart. "O Yes" tells of the growing racial awareness between a black girl and a white girl, friends in childhood but destined to be pulled apart by separate worlds. Some teachers throughout the country have been reluctant to use this demanding, beautifully wrought tale, afraid that its very directness with respect to the question of developing racial awareness in children would turn off many students. But experience has taught teachers otherwise. "O Yes" is neither sentimental nor exhortatory. Olsen wants the reader to understand the decent, idealistic side of children, the bonds that can be forged between them across race and class lines. She is not, in that regard, cynical or gloomy. But she is rather knowing about the manner in which the world's shaping influences bear down hard on even the most eagerly considerate and sensitive. "O Yes" evokes the passion of the black church, but also the social and racial passions of America—the no's that come to our minds as we grow up and learn where we live and why, to whom we belong and with what consequences for our lives. Olsen uses the word "sort" repeatedly; the white girl's older sister says: "They're in junior high, Mother. Don't you know about junior high? How they sort?" She is educating her politically liberal parents, who very much want their children to be in significant touch with black friends. "And you have to watch everything, what you wear and how you wear it and who you eat lunch with and how much homework you do and how you act to the teacher and what you laugh at . . . And run with your crowd."

An Atlanta English teacher at Henry Grady High School told us in 1965: "I divide my class into the pre–'O Yes' stage, and the post–'O Yes' stage. Oh, I divide my own life as a teacher into the pre– and post–'O Yes' periods! I don't know how I ever could have got my students to take a look at themselves, an honest look, without that story. You can ask them to be honest, tell them to be, grade them according to whether they are, but they have their own thoughts and their own habits, and they won't really clue you in to them a lot of the time. They may try to be honest, but they do a lot of holding back. They don't want to talk about what's really going on in their lives—maybe because they don't want to hear themselves say certain things. They spend a lot of time being afraid of one another and trying to earn each other's favor. They can be as tough and sullen as can be, one minute—as independent and willful, also—and the next minute they can be all smiles, and without any mind of their own: all they want is to be accepted by so-and-so, or by the so-and-so's.

"When I read that Olsen book, I was overwhelmed by all that Olsen packs into a hundred pages, four different stories. I hesitated most about 'O Yes'—I thought the students who needed it most would be put off most. They'd tell me: Stop trying to preach to us through your reading assignments. They've said as much a few times, in connection with other reading I've assigned.

"The first time I assigned 'O Yes,' I expected another confrontation when I began the class—asked the class (as I always do) what they thought. Silence. I rephrased the question. Silence. But I could tell it wasn't the silence of angry or irritated or disappointed or confused people. Those kids had been 'hit,' as one girl put it to us. She was the one to get us going. She raised her hand very slowly, and she told me she had a hard time finishing the story. For a moment I thought it was Tillie Olsen's writing that stymied her—the density, the compactness, the tightness of narration. But no, the girl had gotten through all that successfully. She explained herself: she'd been 'hit' by 'all the truth,' and then she added, 'too much truth for one evening.' She had put down the book, and tried to forget it. Sometimes you have to do that, she explained to us—'just put out of your mind what hurts too much to think about.' But she couldn't accomplish that goal, try as she did. She kept stuttering out what she had experienced while reading the night before. It was almost scary, listening to her. A couple of kids were wiping their eyes by the end [of her remarks], though a few others were getting fidgety.

"I decided to be less the teacher than usual. I asked the kids to say what they thought about the story, and one after the other, they did. Then I was ready to move on—but they wouldn't let me. They kept talking. I wasn't even refereeing. They stopped raising their hands. They talked to each other. At one point I was going to say something, but I decided no: keep quiet, I told myself. Be glad they're talking to each other, and to Tillie Olsen. That's what it was—a conversation between twenty-six Atlanta kids, all white, except for three blacks, and Tillie Olsen, wherever in the world she is!"

I tried to find out more, of course, but at first I heard only more of the above. I became a bit more demanding, and eventually heard this: "Those kids were telling each other that their 'clubs' [informal associations in the school] were really 'sad.' They used that word in their own way; I mean, they meant 'sad' as you and I do, but they also meant 'sad' in a morally specific way: a judgment on themselves as cowards, as the timidly ingratiating blind following the assertively fearful blind. After a while the rhetorical questions started to be uttered: 'Why do we behave like this, and why do we act like sheep, and why don't we live up to what we've heard in church, and when will we ever just be ourselves, and stop being afraid of our own shadows, and always trying to figure out what the next guy is saying and doing, or the next girl?' It was the most dramatic class I've ever taught, and the most emotional one, and I swear, I think I can still hear every word spoken, and right now I think I'll never forget every word I heard—those questions, followed by long silences! No one answered any of the questions! They kept asking them, one after the other! I was scared for a while. I think I feared that the class was getting too 'upset.' That's the word that stuck in my mind. And the truth—well, it was I who was upset: I thought I'd failed, because I was afraid someone might say this was no longer an English class in a high school, but some kind of group therapy going on, and so we were way off target—not doing what we were *supposed* to be doing."

The class was not so much upset as powerfully affected by Olsen's capacity to give her readers pause. The same young readers had been able to shrug off lectures and documentary films on prejudice—sociological and psychological statements on the reasons people harbor a dislike of certain others. They gabbed and snickered during the film presentations and while listening to a social scientist do his best to explain the mind's reasons for succumbing to racism. But Tillie Olsen didn't come to them with her finger wagging or with a list of formulations they could readily ignore. Her stories worked their way into the everyday reality of their young lives: watching their mothers iron, and thinking of a story; watching a certain heavy-drinking friend, relative, neighbor, and thinking of a story; watching children in church, and themselves in school, and thinking of a story.

· · ·

The Bluest Eye, a daring novel by Toni Morrison, has unnerved many of my wife's students, black and white alike, and many others of high school age in various parts of the country. Pecola, an eleven-year-old black girl, yearns for a beauty that she believes has eluded her; yearns, too, for a familial happiness that has never graced her life. Her parents fight all the time; her brother runs away. Her father drinks and is violent. Her dream is that somehow she would come to appear different, hence be different —and thus would find herself in possession of an altogether different life. Blue eyes become her obsession, her dream of deliverance. With blue eyes she could say good-bye to the wretched life she knows, could see a new life, a family at peace with itself. But the fulfillment of her vision is denied her. Her drunken father rapes her, impregnates her—and the story ends on that note of bitter disappointment, terrible betrayal.

Not all high school students will want to read this relatively brief novel, but those who have done so report its power. While the author is black and her characters are black and humble people, the story's theme (of the betrayal of innocence) connects Pecola to—well, Portia in Elizabeth Bowen's *Death of the Heart,* a novel that my wife has paired with *The Bluest Eye* to good teaching effect. Portia, who lives in Regent's Park, London, is a white girl of sixteen or so. She is surrounded by wealth and privilege, but slowly loses hope as a jaded, emotionally cold and brittle world closes in on her—hence the novel's title. Portia has those blue eyes, has all that Pecola would ever want and then some. Still, both girls end up spiritually dead. Readers not far removed from them in age, especially young women, have found both those novels hard to forget. Each story presses upon the reader a vision of the adult world as suspect, to say the least, though not necessarily hopeless.

"It's work, hard work, reading those two novels," one of my wife's twelfth-grade students told her—but the student went on to spell out the advantages, in the long run, of such exertion: "I see one television show after the other. I get completely absorbed in them, then I forget them as soon as the next one shows up on the screen. When I read *The Bluest Eye,* I thought I'd have trouble with it, because you have to get into the scene, and it's a pretty sorry one. I don't know any black people, and reading that novel isn't the best way to get to meet a black family! But once I got into the story, I began to forget that Pecola was black, just as it didn't occur to me to think of Portia as white, or even rich, when I was reading about her life. To me, Pecola—well, she was someone who was caught in a terrible jam: a miserable family life. What did she do? She dreamt—I mean, in the day, not only at night. She was trying to escape, and she figured that if she could only *look* different, then things would be different. I hear my mother tell my father: if we could just get away, then I'll come back a new person, and life will be better. I hear her say: I'm going to get that dress; it'll make me a new person. Maybe I'm exaggerating, but my mother has a drinking problem, and she's trying to escape from some troubles she has, and so was Pecola. In my mind, sometimes, I picture myself looking different—sometimes way different. I'll feel a little guilty, as if I wanted to leave myself, say good-bye to my folks and my kid brother, and just disappear. Then I'll catch myself and stop daydreaming. But later I'll be fixing my hair in a new way, or thinking about how I'd look if I went to a doctor and he took out the mole on my left cheek, and I think of Pecola, and I remember when I was eleven, her age, and I'd keep wishing my hair was reddish blond, like some people's. You feel ashamed of yourself for those ideas, until you get to realize that lots and lots of people (maybe everyone) has them—at least sometimes.

"I felt sadder about Portia, to tell the truth, than Pecola. There she was, living a pretty good life. But your life isn't happy because you live in Regent's Park—or here, either [an affluent Massachusetts suburb]. I didn't feel sorry for Portia the way I did for Pecola. I wanted Portia to tell those people off, to run away. That's me: when I get upset, I'm ready to take off. Once I was ready to go further. I packed my suitcase and was all set to disappear. I was going to visit my cousin in California. She's a year older, and very independent. I talked to her on the phone. She said I couldn't stay with her, because our mothers will talk right away. But she has a friend who was going to put me up. But then things quieted down here. I figured I could take it. Soon I'll be away, anyway, in college. After I finished that novel, I kept wondering what kind of person Portia would be when she got older. I guess the answer is that she'd turn into someone like those grown-ups—stiff and cold.

"We talked a lot in class about how you become the person you do become. My mother isn't like her mother. I don't think I'll be like my mother. I like my father a lot, even though he drinks. He's a big, warm, friendly guy. There was no one like him in *Death of the Heart.* It's no fun, being with people who have good manners, good breeding, and no real, earthy goodness to them. But one kid said you get used to the kind of life you're living; gradually you do—so you can't compare how *you* would feel, if you got thrown into a world like that, with how Portia felt. She was already on her way to joining up with all of them, I guess. It's sad, though, to see anyone get lost, and that's what I think happened to Portia—and to Pecola, too. Both the novels make you stop and look around so you don't get lost yourself."

Those excerpts, from weeks of discussion, convey some of what can happen when a high school student begins to think about her own life after giving thought to the lives storytellers tell about. I think it fair to say that my wife's teaching helped matters along considerably. She read aloud portions of those novels and asked for volunteers among her students to do likewise. She showed the class slides of the Regent's Park neighborhood. She connected *The Bluest Eye* to the Olsen story "O Yes," and both of them to the self-discovery her students were experiencing as adolescents. When the students signaled their readiness to broaden the range of reflection, make it more personal, my wife shared some of her own memories—the spells of dissatisfaction with herself, the yearning she felt to break away from the familiar, to explore different ground. Those two young women in the novels, Pecola and Portia, took on a vivid and continuing life for many of the students through my wife's ability to connect their stories to her students' and her own.

A student named Linda, who had read both novels and taken

part in the discussions, came to see my wife after class on a Friday afternoon, a cold, gray March day, to tell her a story. She had written to a nationally syndicated newspaper columnist who specializes in giving advice. She had received a letter back telling her to go seek counseling. The columnist had addressed none of the specific matters she had put into her letter—her worries about herself, her fears. It had taken her a long time to compose her letter, and she had struggled hard with the shame she felt as she saw on paper what she usually tried with all her might to keep out of her mind, only to be overcome by "thoughts"—the most distressing of which, for her, were those in which she felt herself "becoming" various other people: someone she knew in school, or someone she'd only seen on television or in the movies. Linda would "lose" herself, get preoccupied with a person whose looks she liked—or personality, or background, or achievements. She worried that sometimes the person she admired and envied was a girl. She worried that she secretly tried to imitate those people—copy their gestures, their dress, their way of talking. She worried, finally, that she would never stop such flights of fancy, that they bespoke a serious psychological disturbance.

Linda took psychology courses in high school, hoping to learn why she was so driven. She had not been able to confide in anyone about this side of her life—hence the letter to the newspaper columnist. No wonder Toni Morrison's novel struck a nerve in her; no wonder the classroom discussion worked its way into her mind. She read the novel twice, was an eager participant in the classroom exchanges devoted to it, wrote a powerful, touching composition about Pecola's dream—the notion that appearance would triumph over psychological reality—and when she presented her personal story to my wife, made mention of *The Bluest Eye* immediately. The novel, and the classroom talk of it, had given her the first glimmer of hope: it had raised the possibility that she was not all that unusual, and maybe not "sick," but rather that she, like Pecola, had endured some extremely difficult home circumstances and craved some moments of release.

By no means was Linda without what we call "problems." Still, she and my wife were able to address her life's burdens indirectly and directly through Pecola and Portia. She felt she was an amalgam of those two. She was of a well-to-do home. Her father, dying of cancer, had committed suicide. Her mother was always away on "trips," vacations. (Her quite proper grandmother essentially took care of her.) My wife suggested that Linda read other novels of obvious potential significance to her —Nathanael West's *Miss Lonelyhearts* and Willa Cather's *My Ántonia.*

With *Miss Lonelyhearts* Linda could, ironically, make a different kind of leap—put herself into the shoes of someone like the person to whom she'd addressed her forlorn and frightened letter. Linda and my wife had a fine time, teacher and student, imagining the ton of mail brought into the large offices of what (as they both began to realize) must be a major bureaucracy—dozens of people reading thousands of pieces of mail, sorting them out, offering replies. "I'm getting some perspective on *my* troubles by imagining what Ann Landers and her staff must go through every day," Linda told my wife. She could suddenly joke about her "future career goal": to join the tradition of Miss Landers. Linda had hated that expression, "future career goal," because she had not figured out what she might want to do after high school graduation.

Willa Cather's *My Ántonia* gave Linda an opportunity to appreciate the struggles of Ántonia Shimerda, the decent and kind Bohemian girl whose father killed himself; and she and my wife could also discuss the theme Cather wants so insistently to develop—the fatefulness of life, its unpredictable and even mysterious side. Ántonia, too, was a vivid dreamer, anxious to see and taste of life, and anxious to extend herself in many directions. Linda, who had never been near the Nebraska prairie of the late nineteenth century, felt a growing camaraderie as she came to know Ántonia's story. She would eventually tell her teacher that she was beginning "to get lost" in novels rather than people, citing her fascination with *My Ántonia.*

Novels are not, of course, meant to be a replacement for psychotherapy for young people or indeed anyone. But many high school students know in their bones what child psychoanalysts have stated in various ways—that a still growing mind can be exceedingly volatile, can experience wide mood swings, can delight in introspection, can enjoy the pleasures of irony, detachment, satire. It is easy enough to try to formulate such inclinations for those prone to them—that is what the many high school psychology courses in "human development" are all about. Those who teach literature, however, have a different task: to engage a student's growing intelligence and any number of tempestuous emotions with the line of a story in such a way that the reader's imagination gets absorbed into the novelist's. My wife would eventually hear Linda describe such an achievement: "If you stop every few chapters, and think about what the characters are doing, and why the novel is going in the direction it's going, and if you try to think of yourself as first one character, then another, then you'll be close to the story, the way the writer wants you to be. I hate myself when I have a 'fast read.' I get competitive, or I want to get an assignment out of the way, so I just breeze through the pages, getting the gist of the story, so I can say, 'I've read it.' Then you listen to the teacher say something clever about the novel, and read in some 'trot' what it's supposed to be about, and pretty soon, you're through with it, and you can go on to the next one!

"I think it's the troubles I've had that have stopped me in my tracks sometimes—I mean, they make it harder for me to forget a lot of what I read, if I do the reading halfway carefully. And the teacher—she wouldn't let us plow through the novel. She kept saying she didn't even want to test us on the books; she just wanted us to 'take the books to heart.' What did she mean? [I had asked.] She meant that if you 'live' with the book a while ('Try to let the book settle in and live in your heads,' she'd tell us), then you'll be part of the story, or it'll be part of you, I'm not sure which. Maybe both. She'll take a dull old novel like *Silas Marner* (no one wants to read it, and you wonder why it's on any high school list), and she'll start in with it, teaching it, and soon you're not in some English town a hundred years ago, with an old hermit and his gold and the girl he's got to take care of, you're here in America, and it's your hometown, and you're thinking about people who take advantage of others, and people who become victims, and how sometimes if you've gotten a raw deal, you turn to money, to hoarding possessions; or maybe you're like that because you're not a very good person, but then something happens in your life, and you change. You can't give up on life, and you shouldn't give up on yourself. But when I talk like that, I'm sounding like the 'advice to the lovelorn' writer in the paper I wrote. I don't know how to say what happens when you read a good story: it's not TV and it's not reading the paper. It's not the movies, because you get into them faster, but you're 'out' real fast: you forget what you've seen, because the next flick has come, and you're looking at it. With a novel, if the teacher holds you back and makes sure you take things slowly and get your head connected to what you're reading, then (how do I say it?) the story becomes yours. No, I don't mean 'your story'; I mean you have imagined what those people look like, and how they speak the words in the book, and how they move around, and so you and the writer are in cahoots."

A testimony of experienced collaboration, with a teacher as intermediary; a love encouraged by someone whose job it is, actually, to do just that. It is astonishing how young Americans in the late 1980s, some from "culturally disadvantaged" homes, can take to a novel such as *Silas Marner*— which is all too often written off as belonging to an old-fashioned high school curriculum. The old man Silas, so readily regarded as a hermit and a miser, is, of course, a victim of fate, of chance and circumstance. The novel treats of his moral and spiritual transformation while exploring the hypocrisies and duplicities that his society has learned to ignore. It is refreshing and exhilarating to see young men and women in a ghetto high school feel "in cahoots" with Silas, feel his isolation, his bad luck, and react with anger at the wrongs done him. It can be instructive, also, to witness George

Eliot's shrewd social knowledge being taken to heart—her capacity to size up the various classes, give narrative life to their habits, their preferences, their distinct experiences. In ghetto high schools or in private schools that cater, mostly, to rather privileged families, many novels of the nineteenth and twentieth centuries can elicit a broad range of empathic responses.

Linda reminded us that she was just "starting out"; she had years to go, she hoped, and more stories to read. When *Silas Marner* appeared on TV, a two-hour adaptation by Masterpiece Theatre on public television, she had quite a time fighting at home for her right to watch that program. But she persisted, despite the family's initial scorn for both the channel and the program it was showing. Her older brother was not to be persuaded, but her younger sister, twelve, reluctantly decided to give the program a try, and the result was edifying for her and for Linda, who felt herself becoming a teacher. But Linda had an interesting lament. She said she was glad she'd read *Silas Marner* first, and sad that her sister had come across the television version before reading the book. She was not sure her little sister would ever read it.

George Eliot's town of Raveloe, Mark Twain's Mississippi River town, Willa Cather's town of Black Hawk on the Nebraska prairie—these are hardly places that seem suited for today's young folk. They are towns that belong to books not usually judged "relevant." Moreover, *Huckleberry Finn* has stirred its fair share of controversy, even though my wife's black students have not been offended by the portrait of Jim. Quite the contrary: they see his moral superiority to the men and women who inhabit the river towns. And they identify with Silas Marner, right off, and move easily from comments about the rich village squire Cass and his son Dunston, a thief and a liar, to more general comments about aspects of the social order to which their families have belonged. What happens, wondrously, is the kind of moral communion that Linda described: a writer's scrutinizing and suggestive images catch hold of a reader's impressionable, yielding sensibility, and thus they become "in cahoots." No wonder she had her very own idea of what Silas Marner and others in the novel looked like, and consequently found the television presentation a bit jarring. She wanted to keep her own visions, not have them give way to those appearing on the nineteen-inch screen in her living room. Still, she was grateful to have been put in touch with another version of *Silas Marner*, and even had the extraordinary good sense to remind herself, and later her English teacher, that "there are lots of Silas Marners," meaning that the novel inspires in each reader a different series of thoughts, memories, images—and not always those the teacher sees.

·3·

Finding a Direction

In the last chapter of *Middlemarch,* George Eliot asks a question many college teachers have put to themselves: "Who can quit young lives after being long in company with them, and not desire to know what befell them in their after-years?" I have found myself asking the same question as I say good-bye to the dozen students who take part each year in a freshman seminar I teach on the literary documentary tradition, with emphasis on the writing of James Agee and George Orwell. As George Eliot wrote, "The fragment of a life, however typical, is not the sample of an even web: promises may not be kept, and an ardent outset may be followed by declension; latent powers may find their long-awaited opportunity; a past error may urge a grand retrieval." We who work in college classrooms watch as students try haphazardly or earnestly to figure out where they are headed in the time that follows graduation.

Most campuses now offer "counselors" of various kinds. For those who are looking ahead with excessive fear or confusion, there are doctors. For those more moderately perplexed, or simply quite curious and a bit unfocused in their contemplation, there are advisers and tutors who have facts to offer, information about requirements, programs, possibilities, rewards. Many students make their decisions gradually, or by fits and starts, or much to their own surprise, or all of a sudden—but, in any case, essentially on their own. During those stretches of inward looking, novelists may be summoned in daydreams and reveries, or consulted directly, through recourse to a particular page, chapter, section, of a particular story. We all remember in our own lives times when a book has become for us a signpost, a continuing presence in our lives. Novels lend themselves to such purposes; their plots offer a psychological or moral journey, with impasses and breakthroughs, with decisions made and destinations achieved—though it often remains a mystery to me, despite my years of teaching, why a given student has chosen *that* novel (rather than one I might have predicted as more congenial) as the one to take a decisive role in shaping a future life course. The college teacher, needless to say, is not without influence: his or her way of using a text can prompt more than one student to embrace it. It is also fortunately true, however, that even the most uninspired teaching cannot always succeed in keeping one or another student from making vigorous and persisting use of a short story, a novel.

I well remember a student from eastern North Carolina who had stumbled upon Ralph Ellison's *Invisible Man* almost by accident—saw it in a friend's room. It was the title that caught Ben's eye. A junior, he had for some time regarded himself as a bit of a failure, someone who seemed to fall between the cracks. He had been unable to imagine himself pursuing any pre-professional course of study, had even found it difficult to choose a field of concentration. He had also found himself ignored or rebuffed by those he liked. As he once put it, in retrospect, if he had disappeared, "no one would have noticed."

It is not surprising, then, that the title caught Ben's attention. He picked up *Invisible Man,* leafed through its pages, and put it down, unsure what it was about. A friend gave Ben a not original or compelling précis of the novel, and that seemed to be that—one more fact heard but not necessarily retained. Yet the title had its own staying power, as the twenty-year-old student soon realized. The words "invisible man" kept coming to mind. At first he didn't make the connection between those words and a novel's title. He thought, rather, that with his penchant for spy stories, for television mysteries, he had in some fashion craved the symbolism of those two words, the fantasies his mind attached to them: "I'd think of 'invisible man' and I'd be him—someone who could solve crimes or spy successfully because no one could see him. My father used to tell me about a radio program called 'The Shadow'; it was on on Sunday afternoons, and there was this guy, Lamont Cranston, wealthy man about town, and he had a girlfriend, Margo Lane. No one knew it, not even she (I think!), but every so often, when some criminal was on the loose, Lamont Cranston turned into 'The Shadow.' He could see others, but they couldn't see him. He could corner thieves, catch crooks, frighten the daylights out of killers. My dad bought some tapes of old 'Shadow' programs and played them for me; and frankly, I thought the thing was corny at first, but the idea, that stuck with me. And I loved some of the smooth, clever dialogue between Cranston and his girlfriend.

"I kept thinking of 'The Shadow' when the words 'invisible man' came to my mind, until one day I saw another friend of mine with that book on top of his spiral notebook, and I suddenly realized I'd got that phrase from the title of the book. So I decided to go buy it and read it. I'm a slow reader, and I had trouble at first getting into the novel, but I did, and once I was with Ellison, I stayed with him; I mean, I 'connected' with the invisible man. I think, after a while, I began to see people the way he did: I watched people and tried to figure them out. I didn't want to be an outsider, but I was—the way black people are for us, for lots of us. The more I looked at people through my outsider's eyes, the more I felt alone and ignored; it was no fun—and it made me understand not only my own social problems, my trouble getting along with people, but how black people must feel in this world, especially when they come to a place like this."

Those were not especially original or profound comments, nor were they meant to be. They were the impromptu remarks of a student not majoring in English or the humanities—someone taking mostly science courses and headed (he thought then) for a career in geology, engineering, or what he called "the air space industry." Ben's personal isolation—a certain awkwardness of manner, a lack, actually, of the kind of sophistication Lamont Cranston possessed in abundance—had fueled in him one of Walter Mitty's fantasies: the detective or spy who wins everyone's admiration and praise. These fantasies, in turn, had made him attentive to Ralph Ellison's title, and soon enough he had read his first novel in a long time, while telling his roommates that for a while he'd thought he could go right through college without reading a single novel.

I got to know Ben because he took one of my undergraduate courses, one that offers little but novels to the students. He came to see me during my office hours; he wanted to talk about the Ellison novel well before the time I was to lecture on it. He kept coming back, kept wanting to understand the subtleties and nuances of the novel, kept revealing to me—slowly but unmistakably—his serious involvement with the story, with its nameless protagonist. At times, I have to admit, I found his statements naive, inappropriate, extravagant. I didn't like, for instance, the way he worked so hard to connect "The Shadow" to *Invisible Man.* Nor did I like the extent to which he kept forcing similarities between his college life and that of Ellison's main character. On the other hand, he was devoting a lot of time to one book, reading it as carefully as he knew how, rereading it, going to the library to search out critical essays on the novel. He was educating himself, really, as a reader and critic through the access he gave himself to one novelist's story. By the time I did give my "Ellison lecture," I felt considerably in this student's debt. He had prompted me to take the novel more seriously myself. Before, it had been an important book in a list of important books; now, it was a book that a student of mine had been calling to his side in an intensely personal way. Moreover, I was stunned at how serious Ben could be as he contemplated a fictional person's invisibility, and thereafter that of others—blacks especially, but also some white people near at hand.

One day Ben told me that he'd just, for the first time, noticed and spoken to a man he passed several times almost every day:

the person who checked students as they entered the library (for a proper identification) and as they left it (for what they were carrying out). Why is it, the student asked, that he had no idea who this man was, what he looked like, though he'd seen him daily? Many times the man had spoken to him, he began to realize, yet he had never replied even in brief kind: "I first noticed him, *saw* him, a few days ago. I first *heard* him then, too. The guy ahead of me was checking out. That man thanked him for opening up his book bag, and smiled at him. He told the guy to have a good day—I guess because the student looked at him. You can acknowledge someone through a look. You can dismiss someone completely by not looking at him. That's what I'd been doing. I never paid attention to him.

"That afternoon, I realized what a rude son-of-a-bitch I am. I realized that I'm full of myself; I have no time for anyone else, unless it suits my own needs."

He tried to make amends, of course, when his turn came, right after that of the student he'd observed having a brief, friendly time with the balding, middle-aged man, who was (it happens) black. He looked right at the man, caught his eyes with his own eyes, smiled at him and received his smile, listened to the man's words as he thanked Ben and wished him a nice day. Ben responded: "And you, too."

"I knew I shouldn't overdo things," Ben later told me, "but each time I got a bit friendlier with him. Soon we were chit-chatting. I'd tell him it had started raining; he'd tell me he was glad he brought his raincoat—talk like that. He *recognized* me. When I was in line, he'd smile at me while checking the books of someone ahead of me. When it would be my turn, he'd smile again and we'd have our exchange. It meant something to me, those words he spoke and the ones I did. Finally, one afternoon, he broke the ice further; he asked me what I was concentrating in [what his major was]. We had a real nice conversation."

Such a conversation, in all its brevity yet significance, could well have been out of *Invisible Man* itself, as Ben realized in retrospect. Ellison is a master of rendering the rituals and routines, the gestures, the shorthand signals, that enable people to come to terms with each other across the barriers of class and race. Ellison's hero has traveled a long, complex journey from the rural South to Harlem, along the way shedding innocence and acquiring savvy, even as this college student was learning how to comprehend a library guard's mix of polite circumspection and shrewd friendliness. "He doesn't miss a trick," Ben observed once. The guard watched the faces of the students, took their measure, knew in an instant who wanted to speak with him and who had no such interest. By the same token, the guard knew with respect to each student how far to go—with a smile, with words, with a tilt of his body or a wave of his hand or a modulation of his voice. By then Ben, taking a course on the history of jazz, knew that Ellison was thoroughly conversant with jazz, too. With the language of jazz in his mind, Ben described the guard as a "solo musician," someone who possessed a lyrical, swinging kind of empathy. The guard became a significant teacher, even as *Invisible Man* became an important moral companion.

Ben tried to test his new understanding by putting himself in odd situations. Once, having a paper for my course to write, he went to a hotel lobby, sat down at a desk there, and began writing. He had with him a copy of *Invisible Man,* which he put on the desk along with sheets of paper and a pencil. He sat there—he was the only person sitting in the lobby—and watched people; watched the hotel staff answering the questions of guests, rendering them services of various kinds; and watched, eventually, an apprehensive functionary's growing interest in him. Ben realized it was only a matter of time before he would be confronted. "I sat there waiting, and finally he came up to me and asked me if he could help me. I said no, no sir. Was I waiting for someone? No. Was I writing a message for someone in the hotel? No. (I should have lied, told him I was writing the story of my life, and was going to leave it in someone's box—someone staying in the hotel.) His eyes swept from my jeans and old, torn sweater to my book, to my pencil and paper, and he said I was in a hotel lobby, not a library or a dorm common room, and the desk was not to be 'monopolized' by anyone for 'too long.' I could have fought with him. I could have dug my heels in. But I left. I kept thinking of Ellison and his novel—how uptight people become with other people, how they won't really see those people for who they are. It's a kind of blindness we develop. You can't have someone invisible without there being someone who is blind!"

He tried to imagine, too, how he might have been treated—at that time, in that place—had he been black. He also began to put himself in the place of the homeless. One weekend, again intent on drawing on a personal experience, he took off—left his dormitory for the streets of the city. He walked and walked. He panhandled. He tried to sleep on the grates outside an apartment building, the steam enveloping him, weakening him. He went into a restaurant and begged for some throwaway food. He relieved himself in the dark corners of back streets. He felt his beard growing, felt his mouth becoming more and more stale, felt his stomach hurting, his back and feet, too. He felt a fierce, piercing headache. He reached for some food in a garbage pail outside a restaurant and sensed his gorge rising in protest. The dry heaves were an expression, he realized, of his lingering bourgeois values. He wondered what in the world he was doing—was he showing off, playing the fool, turning the misery of others into an excuse for his own spell of self-indulgence? Not least, he became weaker, more racked by muscle pains, by an insomnia that wouldn't yield to the strongest desire he'd ever felt to fall asleep and stay asleep.

After forty-eight hours of the self-imposed turmoil, he surrendered. That was the word he used: "I felt I was losing my mind—that's when I surrendered. As bad as the troubles my body was going through [turned out to be], it was my mind that did me in. I was jittery all the time, and when I saw someone coming near me, just walking on the sidewalk while I was walking, or while I was sitting down on a curb, I thought they were out to get me, that they'd beat me up or something. I knew lots of street people get robbed, and so I took no money with me. I figured I'd be tempted to bow out, to go fill my stomach and maybe find a room for the night, a cheap place. I didn't even take my [room] keys with me; I was afraid I'd lose them. But one thing I forgot: I left my watch on my left wrist! It's a Rolex. For a while, everyone had spotted my Rolex. Everyone was out to get me in some corner, and take it. Then I'd be really 'down and out.' "

Abruptly, in the middle of his second homeless night, he found himself peering, with tears in his eyes, at the illuminated dial and hands of his Rolex. He took it off, hearing as he did so the snapping noise of the bracelet as if it were an old familiar home noise. He began to cry. He noticed the bad taste in his mouth. His nose was dry. A dull ache had settled on his stomach. His hands were dark with dirt. He pictured himself in a hot shower, standing there for an hour. He imagined himself in his bed, sinking into the soft mattress, his head on the pillows he knew so well. He had taken all that—the everyday particulars of his student life—for granted.

As he stared at his watch, observing the longer hand move from quarter past three to half past, his mind was suddenly awake. He was no longer silently crying. Words from books echoed in his head, which was both heavy from fatigue and light from the effects of hunger. Ellison's "invisible man" appeared first: "I saw him! He wasn't invisible to me. I imagined him—I mean, I pictured him. He was talking to that college president in Alabama; then he was in Harlem, standing on a street corner, looking at all the black people go by, some pretty funny folks and some plain, hard-working ones, doing their damn best to get by. I tried to imagine him in that paint factory, and talking to that Harvard type, that Yankee philanthropist. I pictured him in Harlem, staying with Mary Rambo. I pictured her, too. I wasn't far from a black [neighborhood housing] project, you know. I'd seen some black kids, and thought of Ellison and all he was trying to say.

"I was on thin ice, maybe—in my head. I ran through your reading list. I wondered what Orwell would make of our homeless problem today. Is it worse than what he saw in the 1920s, when he was living in the slums of London and Paris? And what would he think of Jude Fawley? He got to me: the farm boy who had dreams, but they were crushed. When he goes to Oxford and sees all those snobs—it makes you nervous: you think of Harvard."

Ben stopped there in his description of those two days and said nothing for a while. He avoided looking at me. That brief venture had obviously affected him deeply, caused him to take a look at himself, even at his course of study and the place where he was pursuing it. He began to talk again. I was stunned by the connections he was making between his extraordinary experience of the days that had just passed and the books I'd been handing out to students for years. He had thought not only about the books but about the authors' lives. He talked a great deal about Dickens. Dickens had known the harsh life of poverty, even—when he was a child—debtor's prison, with his family. In *Great Expectations,* the boy Pip is elevated by an accident of fate to a much higher station in life. In telling Pip's story Dickens examines the consequences of a rapid social and economic ascent, very similar to what Harvard has offered many of its students over the generations. As Pip leaves his rural, humble origins and takes up (at the expense of an anonymous benefactor) a privileged and sophisticated London life, he is morally challenged. His mind cannot stop noticing the social distance he has traveled. The same was true for Ben, whose mother had once told him to be glad he had ten toes and ten fingers and a head that worked reasonably well. His father had told him to fight for his rights in school but not be a "big shot." What kind of person was that? His father didn't have to spell out the details. Ben knew that good grades were fine, but that the one who tried to push himself all the way to the top was only setting himself up for the envy and resentment of others. Besides, why try to be an academic leader when a factory job was virtually foreordained?

Yet without any great effort this particular child, one of five, had somehow risen to the top in elementary school, in junior high, and in high school, too—almost, it seemed, against his will. He always got the right answers in daily or weekly quizzes. He wrote good compositions naturally. He put down on paper what came to his mind, and the teachers applauded heartily what they read. He was called "bright"; his parents were told he had "promise." They didn't know what to do with such words, nor did he. Moreover, he was a rather handsome person, and soon enough girls took notice of him. For a while he deliberately took satisfaction in that sphere of life—tried hard *not* to do well in school while responding eagerly to the attention of one or another of the prettier girls.

In Ben's first year of high school a teacher saw this happening and decided to pull him aside and try to influence him, push him toward books and away from girls. Ben listened, then said no; but he couldn't get out of his head the idea of great expectations once the teacher had inserted it. He felt, suddenly, a kind of loneliness. Unable to discuss his future with his mother or father or any of his friends, he kept his own counsel. He went secretly to a downtown library and looked at college catalogues.

Eventually he would be a valedictorian, but well before that title became his, college "scouts" began noticing him, Harvard's among them. During his junior year in high school, he hadn't even heard of the place. But as a senior, after various lessons about power, prestige, and money, he sought letters of recommendation, secured money orders for the applications (paid for through several jobs held during school and in the summers), and was interviewed by representatives of various universities. In time the acceptances arrived, and the important decisions had to be made—by himself alone. He had always known that such decisions were his to make. Not even the teachers who had encouraged him were consulted. On a long walk his mind turned northward, crossed the Mason-Dixon line, passed through Pennsylvania and New York, entered New England, and settled upon the oldest of the colleges. Harvard had offered him a scholarship. "A good deal" was the phrase that kept entering his mind as he stared at the Carolina countryside.

Ben hadn't paid those earlier years much heed after his arrival in Cambridge, Massachusetts. Quite the contrary: he had resolutely put his Dixie life out of his mind. As he spoke the word "Dixie," playfully and insistently, in my office, a whole world returned, not only its sights and sounds but its values and expectations. Suddenly, half jokingly, he announced he had been "forgetting" the full title of the Dickens novel—he had once absent-mindedly called it "The Great White Hope." How revealing, he declared. Harvard had become the "great white hope" of this young man from a self-consciously, indeed militantly, white, working-class background. He and Harvard had become "co-conspirators," he declared.

In explaining that last word, he began to talk at length about loyalty and betrayal, on the changes, moral and psychological, that college life had wrought—and how swiftly! Clothes, preferences in food, in movies, in television. He had gained a sense of personal security—and lost, he now believed, no small amount of self-respect. Another fantasy entered his head: a sign or poster, fashioned in red crayon or paint, proclaiming to all who entered his bedroom in the snobbish Harvard dormitory, DICKENS AND PIP LIVE! But he knew that what really counted in the coming year, before he was "sprung" into the outer world, was his "attitude." He hesitated to expand on that word. I resorted to the usual questions and entreaties. But Ben turned silent and, I thought, melancholy. He seemed to change the subject: the Atlanta Braves were still his favorite; he would never become a Red Sox fan, come what may! But he hadn't changed the subject after all—so he realized—and gave me a darting glance.

Now silence took over, held its own for an uncomfortably long time. I was preparing to find an excuse that would enable us to go our separate ways when suddenly Ben began an extended monologue on the novel *Great Expectations* and on George Eliot's *Middlemarch,* and especially on Dr. Lydgate in *Middlemarch.* When he talked about Pip, he was a bit repetitious, though his moral passion was compelling and contagious. When he talked about Dr. Lydgate, however, he was the vigorous judge of a book he'd carefully read, and was also the watchful critic of his professor. In time, that latter aspect of his "attitude" gathered a great deal of energy, and my wish to end our conversation decidedly increased. I protected myself with a familiar strategy—I told myself that this young man (now in my mind a mere boy) certainly had his problems, hence his need to turn on me.

Ben was not only evoking the moral turmoil of Dr. Lydgate but reminding me how that character in *Middlemarch* was meant to challenge other doctors. Had I changed with respect to my goals or purposes as a physician in the course of my life, and if so, how and why? True, he quickly withdrew—apologized, that is, for what might be regarded as an unwarranted act of boldness. But he knew that I was not easily going to dodge his moral inquiry, especially in view of his long spell of self-scrutiny. I answered, finally, by telling him stories of my high school and college and medical school days, stories that responded to the thrust of his comments about what happens to us as our self-expectations are confronted by new realities in our lives.

One of the things I told him was that I had once wished to work as a public health physician in a needy rural American community, or maybe abroad. He let me know that he had earnestly "back then" pictured himself, in years to come, as a Southern high school teacher, reaching out to others as a teacher or two had done to him. He had abandoned such an ambition—not on its merits, he hastened to tell me, but out of a strongly felt shame. He could not think of mentioning to anyone he knew that he envisioned such a return to native grounds. He recognized his reasons. He had, in fact, heard others (not many, alas) speak of their future lives as teachers, as political activists. But he doubted their sincerity even as he himself set his sights on a high-powered Yankee life he had only begun to comprehend in his sophomore year at college.

During other exchanges in the weeks that followed, those characters of Hardy, Dickens, Eliot, inhabitants of another century, came increasingly alive. They seemed to speak not only for themselves but to one another. Jude's disenchantment with the intellectual life he witnessed in Christminster (Oxford) was brought to bear on Pip, who had become all too entranced with London's upper-class life. George Eliot's shrewdly acerbic comments ("the Christian carnivora") were directed at the modern secular pieties of a self-important college community. Both of us tried to grapple with what all three of those novelists were stressing—the influence constantly exerted by social class, by the nature of everyday circumstances. Moreover, Ben kept reminding me as well as himself, so much of "all that" happens "unconsciously." Had I ever realized that to be the case, for all my professional interest in the workings of the mind? I had to acknowledge that I was no less an unwitting product of my social and economic situation than he! But he would have no part of my excuses. He asked aloud whether my profession, so intent on having me look inward, had ever insisted I look at myself closely in the fashion he and I had been doing of late—in the manner, really, Eliot, Dickens, and Hardy had asked us to do. Since the answer, once more, was no, he was able to suggest that perhaps those writers had something to say to teachers as well as students—including teachers who analyze texts closely, all the while skirting or ignoring entirely what he had heard a lecturer call the "moral intention" of the author whose novel was under examination.

He reminded me that Dorothea Brooke's "theoretic" mind had not always helped her understand people. He reminded me that many novelists—Dickens, Hardy, Eliot, Tolstoy, Dostoevsky—were in their own separate ways notably distrustful of intellectuals. He wasn't personally truculent toward me, or self-disparaging, either, but he was intent on exploring a distinct irony: "I think it's strange, the call to arms of some of these great writers. They're not as impressed with high-and-mighty minds as this place [Harvard] is; they're even ready to turn to poor and not so educated people for moral guidance. What does that tell us about ourselves, and why don't we take their message to heart? We get graded on how well we understand their message. But there are times when I think the entire exercise is pretty strange. I get an A on a paper about poor Jude Fawley or Dr. Lydgate's moral decline, when the point is that I am exactly like those Oxford students who ignored Jude, and I'm probably headed in Lydgate's direction myself. Why don't all of us—the teachers and the students—try to take these books to *heart,* not just analyze them and then go on to the next book? We may be smarter, but are we better?"

I wanted to know what he meant by "better." I damn well did know, but I didn't have the courage or heart to face up to his moral challenge to those of us who teach one another, who learn from one another, through the agency of books. Ben became impatient with me. He let me know that I was dodging a bit, that he didn't want to get into an analysis of what makes for moral virtue. After all, a thorough Oxonian knowledge of moral virtue on the part of all those bright students helped Jude Fawley very little.

Ought Jude haunt us in Cambridge, Massachusetts, in this century? Ben was becoming rhetorical, I noted—which made me turn cold and analytic. In fact, I was puzzled and stymied. A student was posing an important dilemma. I was questioning him, fending him off rather than joining him in an acknowledgment of a genuine moral difficulty. Finally, he made an effort to let me off the hook. "I guess all we can do is talk like we've been talking," he offered. Then he admitted a keen disappointment: "I wish someone would figure out a way to get those novels to stick to us, so that we don't forget them a day or two after we've read them, or right after we've taken an exam, and written our clever essay on some topic the professor has given us to consider, like: 'Discuss Jude's attitude toward education.' I don't have the courage to tackle questions like that the way they should be tackled. I mean, why not discuss places like Oxford and Harvard? What is *our* 'attitude toward education'? Hardy took aim at academic snobbery, and it would be great if we were asked to connect what Hardy wrote with what we see now. What's the point of reading *Jude the Obscure* if you don't stop and ask yourself about the Judes out there beyond the Harvard Yard, who might feel about us the way Jude felt when he came to Christminster? I guess it's too painful for all of us to think that Jude might still be alive."

He was even more sardonic: "I can't believe Hardy didn't want the Oxford and Cambridge teachers and students who read *Jude* to take his novel to heart—to be made nervous by reading certain parts of it [the Christminster section], to be made *ashamed,* not only of what used to happen, but what is still happening. One way to read a novel like that is to see it as a challenge to your conscience, not just your intellect. Aren't we here to grow a little in that direction—to become self-critical as well as critical?" A long pause, and then he told me: "You don't have to answer that one."

I was dead silent in my chair, trying to frame a reply. I had it within me—agreement. Ben's question was rhetorical, and no one would really argue against the desirability of a scrupulous examination of one's own motives and purposes. Still, I wondered how a college is to involve itself in such an effort. I had heard a distinguished professor make the point that "universities offer courses, whereas students are the ones who have the freedom to do with those courses what they wish," an interesting distinction. Here was a student, however, who seemed to be asking for a kind of personal expedition, a kind of introspective inquiry that his professors had yet to offer. On his own he had been reading William James and Tolstoy. As for the books used in my course, he had read them carefully, taken them quite personally. I found myself at a loss for any suggestion other than that we who teach college might want to ask students to keep a journal in which their private reflections and experiences keep company with the moral thinking various writers or (through their lectures) professors might have prompted. I suggested as much to my young friend, but he was not terribly impressed or mollified.

He was afraid, he said, that the writing would become an "intellectual exercise." I grew impatient. That's what colleges are *for*— to help students learn to perform intellectual exercises. Colleges are not in the direct business of telling students how to live their lives (though of course what students read will, one hopes and prays, have a decided impact). This was not new territory we were exploring, we both knew, but I understood this student's intent: Ben wanted *Jude the Obscure* and *Great Expectations* and *Middlemarch* to be taught in such a way that they were not only nineteenth-century classics but urgent commentaries on twentieth-century life.

. . .

Some novelists, of course, are forthrightly concerned with ethical reflection. Tolstoy and Dostoevsky had no reluctance to place religious and philosophical questions in the center of their plots. Contemporary American fiction writers such as Walker Percy and Flannery O'Connor address matters of the soul as they construct narratives. O'Connor's favorite of her twenty-four published stories was "The Artificial Nigger," a tale at once comic and gravely serious.

In "The Artificial Nigger" Mr. Head (a white man) takes his grandson Nelson from a rural east Tennessee world to the big city, presumably Atlanta, for the day. On the train that they board in the early, dark hours of the morning, the old man is at pains not only to educate his young charge but to display his own elderly worldliness. Much of the story is about that process: innocence slowly being shed under the guidance of a determined teacher. But the author has more in mind, and tips her hand in the name she gives the old man. This Mr. Head is a smart one, knowledgeable and determined to make it clear how much information he has stored, how heady he is. Their experiences in the city, as the grandparental teacher hurries his grandson onward, descend into a hellish estrangement. Mr. Head's self-importance, his pretentiousness, his cowardice, become steadily more evident as the story moves along.

In a terrible, climactic moment Mr. Head even denies knowing Nelson rather than be publicly identified as related to him. The incident leading to the denial is a trivial and funny one—a woman with groceries gets hit accidentally by the boy, who has awakened from a brief sleep and runs in panic out of fear that he is alone. (His grandfather is out of sight, hiding.) Some people standing nearby frown with displeasure, prompting Mr. Head to dissociate himself from his grandson. The deed proves devastating to Nelson. He loses faith in Mr. Head, whose moral character has been exposed: he is a person who would deny his own kin for fear of a moment's public embarrassment. Earlier in the story Mr. Head asked the boy whether he'd ever "seen [him] lost." The boy had not. But now they both were "lost," estranged from each other, "east of Eden." Their reconciliation is as ironic and chancy and gratuitous as life itself can be. As they walk the city, which has become a desert for both of them, they come upon "the plaster figure of a Negro sitting bent over on a low yellow brick fence that curved around a wide lawn." O'Connor's description of that figure gives the reader some moral direction: "It was not possible to tell if the artificial Negro were meant to be young or old; he looked too miserable to be either. He was meant to look happy because his mouth was stretched up at the corners but the chipped eye and the angle he was cocked at gave him a wild look of misery instead." O'Connor's description of the onlookers provides additional moral force: "The two of them stood there with their necks forward at almost the same angle and their shoulders curved in almost exactly the same way and their hands trembling identically in their pockets. Mr. Head looked like an ancient child and Nelson like a miniature old man. They stood gazing at the artificial Negro as if they were faced with some great mystery, some monument to another's victory that brought them together in their common defeat."

The story is about the humbling of Mr. Head; intellect's swollen pride is laid low by an author keenly Biblical in her sensibility. The story moves thematically from that initial pride to confusion, isolation, despair—and then to a moment of grace, which brings redemptive reconciliation. The secular reader need not subscribe to Flannery O'Connor's religious outlook in order to gain a moment of ethical instruction from a tale that is, at the same time, arresting and humorous in its narrative presentation. The ways in which a cold, self-serving, arrogant, smug intellect can betray others—and ultimately itself—are numerous and have had, in their sum, a large place in recent world history, as students themselves discover when reading such books as *The Best and the Brightest* and *The Nazi Doctors.* Those two books document moments in the twentieth century when the callousness and destructiveness of well-educated people showed themselves: one book examines the behavior of our foreign policy establishment during the escalating Vietnam War; the other, the complicity of German physicians (and members of the intelligentsia) with the Nazis. Flannery O'Connor took deadly aim at the mind's reasoning powers—how they can be used, to what they can lend themselves. It is precisely such matters that many students are eager to discuss if given permission to do so.

Once, as a number of graduate assistants who taught the sections of my undergraduate course sat and described the reactions various students demonstrated to the stories we teachers had assigned, one delivered a soliloquy on Flannery O'Connor, of provincial Milledgeville, Georgia, and on her astute readers, who were at least temporarily living in the cosmopolitan city of Cambridge, Massachusetts: "At first many of the students have a lot of trouble with her. They want to know what she's getting at. They find her dense, though intriguing. They are fascinated by her cleverness—and tricked by it, and, in the end, exposed by it. She catches the intellectual reader in the same trap that she springs for the intellectuals in her stories. It's probably the same trap she wants to set for herself! I guess I'm assuming that she wouldn't spare herself the derisive satire she levels at her heroes and heroines—Hulga in 'Good Country People,' and Julian in 'Everything That Rises Must Converge,' and Sheppard in 'The Lame Shall Enter First.' They're all the same really; they're all as heady as our friend Mr. Head, and as cut off as he is from goodness. They lack kindness and thoughtfulness toward others, whether you call those God-given qualities or not. At first the students are as heady as Mr. Head is in the beginning of 'The Artificial Nigger.' Like him, they are ready to show off—let any Nelson know that they are as smart as can be, as clever at interpretation as the admissions committee expected. That's what section meetings in Harvard courses are made up of: cognoscenti teachers and cognoscenti students! But in her stories O'Connor turns the tables on you—on us, on me!—at a certain point, usually near the end. I find, actually, that it's best to tell the students to read the stories twice, and I anticipate their troubles by warning them of surprises, of difficulties ahead. Of course, they then want an immediate explanation! But nothing doing. I tell them simply to be careful, because the author has planted traps, and—the voice of experience—they might get caught.

"They start in first, the showy and clever and talkative ones, then the quieter ones who disagree or, often, are the ones who have begun to catch on. I'll never forget the moment in one class when the quietest one of all—he had never opened his mouth—suddenly spoke without raising his hand. He said he thought we were all being Mr. Heads, himself included—the way we were trying to prove to the world how 'high-and-mighty smart' we are. Boy, do I remember that phrase, and his Virginia accent, which sounded like God's pronouncement at that moment, because of the subject matter under discussion! After he'd spoken, the class exploded. No one was angry at him, really. He'd clued his neighbors in; he'd got them off one track, and then they followed another track. They became more personal, more confessional. They connected O'Connor's stories to memories of theirs, to experiences they were having right then and there in their lives. Well, this gave me my chance to discuss what I believed to be O'Connor's purpose: sure, to tell a good story in a persuasive manner, but also to reach some moral and spiritual side of the reader."

Another section leader treated us to the recollection he offered his class: "I told them of my Boy Scout days. I was really competitive. I accumulated a mountain of badges. I became an Eagle Scout. I would push and shove my way through any hurdle or ordeal; I wanted to win, to be first, to get more recognition than anyone I knew. By the time I was sixteen or seventeen I was a big shot. The younger kids looked up to me: boy, did they! The kids my age got out of the way, ran for cover. The other Eagle Scouts greeted me as a 'comer.' Proud—was I proud! I loved having admirers, and the younger the kids, the more glory I got. The Cub Scouts thought I was a living hero! They'd crowd around me and ask me questions, and I'd smile, a big smile, and pretend to be just another guy, and I'd answer their questions.

"I told the students what happened one day. I was leading a bunch of twelve-year-old boys up a hill, along a Scout trail. It was marked, but you had to keep your eyes open or you'd miss the signs—yellow dots on trees—and get lost. I knew every sign, of course! Some of the kids were getting tired, because the trail was a steep one, and we'd been going all morning, with only a short stop to drink water from our canteens. They tried not to let on; they kept walking, but they were going at a slower pace, and I had to exhort them, once or twice, with 'Faster, please.' One kid even started lagging way behind, and when I went back and asked him what was wrong, he said, 'Nothing,' but he looked really beat to me, and I didn't reprimand him. In fact, I began to think we should turn around and go back down the hill. But my own pride, my competitiveness, wasn't going to go along with that idea. I kept leading the boys, but my mind was elsewhere: I was worried about that straggler, and I was trying hard not to lose my control and get angry at the boys for not moving as fast as I thought they should be moving. Then I noticed the straggler was falling even further behind. I stopped everyone. I went back to the straggler and asked him what was up. He said, 'Nothing.' I said, 'Baloney.' I asked him to be straight with me. He said he *was* being straight with me. I heard myself say 'Baloney' again. He looked at me but said nothing. I raised my voice now: 'Are you sick, tell me!' He looked at me and said, 'No, honestly, no.' I shot back: 'Then you're tired, right?' He said, 'So what!' There was something about the way he said it that got to me. I fell apart. I shouted at him; I called him every lousy swear word I knew. He just looked at me, and then I saw a faint smile on his face. I grabbed him. I wanted to know what the hell that smile was all about. I asked him, screaming, to tell me. He didn't. I didn't know what to do. Should I just walk away and take charge of the pack again and get those guys up to the top of that hill? Should I turn us all around and get down as fast as possible?

"By then all the kids had gathered around us. I knew I was on stage. I was standing there, weighing my options, when that kid came close to me, and he said just a few words. 'We'll go; you need us to go.' He was staring at me, right into my eyes. I wanted to choke him. He had that faint smile back on his face. He was testing me. He was teasing me. He'd figured me out.

"That was the first time in my life I'd ever been stopped cold, dead in my tracks. I'd been pushing at the world with all my might. But suddenly all the air was coming out of a big balloon! I felt this urge to cry. I felt I was losing my marbles, as my uncle would say. He was an alcoholic, and he'd tell my daddy he was afraid he was losing his marbles, and I'd ask my daddy what that meant, and he'd always say it meant someone is nervous. I'd want a bigger explanation, but no go. My daddy told me to stop the questions, and I did.

"All that suddenly had come back to me, out in those woods, up that big hill. I was scared—scared of those kids, scared of myself, scared of that one kid. Then he spoke again, and this time he was sweet. He said he wanted to push ahead, and he was sure everyone else wanted to do the same thing, so he moved on up along the path, and the others followed him. Would you believe it, *I* followed *them.* They all led me the rest of the way to the top. To this day, I'll tell you, I can't figure out what happened to them or to me. I just knew in my bones, my muscles, that I shouldn't try to assert my authority during that last half an hour of our climb, that I should just follow them and let things *be.* When we got to the top it was as if nothing unusual had happened. They gathered around me, and we all said 'Great' to each other, and we looked off at the city way beyond and the countryside, and then we were singing and happy, and things were much better."

The teacher's sense of amused detachment and irony had been conveyed to his students along with the tale. The experience must have been similar to that of reading "The Artificial Nigger." Flannery O'Connor's fiction embraces both mordant humor and penetrating seriousness. This teacher's willingness to offer himself as a colleague of O'Connor's Mr. Head, and the young Scout as a version of the boy Nelson, prompted a class of college students to volunteer some of their own pasts for scrutiny. Flannery O'Connor's tales became connected in our minds with some of the hills of pride we had happened to climb.

· · ·

Once in a while a bold student dares really immerse himself or herself in a particular novel or short story, dares give a literary creation a new, private lease on life. One such student of mine, George, vividly described his preoccupation with Hardy's Jude; with Walker Percy's Binx; with the young James Agee, who visited Alabama in 1936; with the young Orwell, who went "down and out" in two European capital cities during the late 1920s; with the writing doctor William Carlos Williams as he did his medical work among New Jersey's poor by day and in the evening wrote his poems, stories, novels; with Ralph Ellison's "invisible man," whose travels and encounters evoke a people's suffering and exile but also its enormous vitality. Like Ben, George was especially taken with Ralph Ellison's protagonist, even though George, too, was white. I was yet again reminded how wonderfully contagious a novel's moral energy can be, at least once in a while. As George put it to me, softly but with obvious intensity of feeling: "I feel I'm 'invisible' as I watch life here go on—or I feel I'm Jude, as confused as he was, staring at people but not part of the world he wants with all his heart to join."

At that point the young man broke off—embarrassed, obviously, by what seemed to be a self-pitying turn in his line of expressed thought. Who was he, George then asked with evident feeling, to "identify" with Ellison's hero, with Hardy's, or for that matter with the young Agee and Orwell? He had lived a lucky, privileged, protected life; had seen little of the hard, rough-and-tumble world Dr. William Carlos Williams evokes, or Dickens, never mind Ralph Ellison or Zora Neale Hurston. (We had read *Their Eyes Were Watching God,* Thurston's powerful evocation of Florida migrant farm life and of a woman's search for her own dignity.) On the other hand, George couldn't help but insist, the men and women he met in those books had become part of his life. He had pictures of them in his mind. Their words came to him at times. He remembered those words; perhaps he would ultimately forget them. But he was not at all sure that the "people" (as he called them) would ever vanish. He still thinks of Huck Finn every once in a while, especially when walking along the Charles and wondering what will show up next in his life, even as each bend in the Mississippi had its fateful persons and situations to offer the traveling, searching, sometimes drifting Huck.

That student, George, was a member of a section in my course taught by Elaine, a young woman who was especially drawn to the Tillie Olsen stories. Elaine taught "I Stand Here Ironing" with special conviction and passion. She emphasized the story's moral introspection: the mother's taking stock of life as it has been lived by herself and her daughter. Elaine shared some of her own family's history with the class. Her grandmother and her mother were poor Irish women who had worked long and hard, hoping that their exertions would enable their children to have at least a slightly better chance at securing a reasonably comfortable life. Her personal memories sparked memories in the students—though she kept returning to the text as the mainstay. "There must be a balance," Elaine once reminded me, and then spelled out its nature: "The stories are emotionally powerful and have a strong effect on the students. They tell me so. When they come to class I want to encourage them to talk directly and without fear or shame about their response to what they read. But I want us to make sure the responses are to the text. Sure, the mind wanders, and to some extent that's all right. But at a certain point, as they talk, I make connections to what we've been reading: that's the teacher's job. The students are usually grateful. They want to digress, but they want to be brought back home, also. Otherwise they feel lost—too much on their own."

She laughed about the imagery, but I had visited her class repeatedly and was impressed with how hard she worked to cover the text, so to speak, while also encouraging the students to let their own experiences connect with those evoked by an author. Her students would remember her teaching.

• • •

George Eliot insists in *Middlemarch* that life can be exceedingly hard to tie down with abstract, categorical formulations, and hard as well to predict. She was constantly reminding her readers, reminding herself, that each person's life has its own nature, spirit, meaning, and rhythm: "Every limit is a beginning as well as an ending," she declared in the finale of *Middlemarch.* What George Eliot knew, many of today's young people know; it is a knowledge that resides in the marrow of their bones, in the chambers of their hearts. College students are forever trying to find a direction for their lives and forever discovering that there are currents and cross-currents to negotiate. Eliot's novels, and those of Dickens, Hardy, Tolstoy, remind them that as in life, so in great art. No wonder the young men and women in my classes are so often anxious to remind themselves and their teachers that the fate of a Binx or a Jude, a Dorothea or a Dr. Lydgate, a Pip or an "invisible man," a Joe or Gurlie Stecher, or even Stecher's creator, Dr. William Carlos Williams (as he is described in his autobiography), is far easier to analyze in retrospect than to comprehend during that fate's unfolding. Again and again, instructed by novelists, students remind themselves of life's contingencies; and in so doing, they take matters of choice and commitment more seriously than they might otherwise have done.

One student, not a good writer but thoroughly sensitive and watchful, was especially eager to read books and anxious to write better about them (he had a mild dyslexia, which cast a decided shadow on a smart and knowing mind). He wrote a paper in which he deplored his own self-consciousness, yet acknowledged how necessary he had begun to realize his reading was, "at least for a while." I wanted to know what he meant by that qualification. He responded with an arresting mix of apprehension and conviction: "The danger when you're young (in college) is that you'll think you're a millionaire of time: you have infinite days and months to spend, and so you can squander a lot of them, and still be way ahead. It's a mistake! That's the big lesson I've learned from reading those novels—that even if you're just marking time, or at the very start of things (in your life), you can get yourself into a groove, or a road that won't be so easy to leave, and that may take you a long way up or down, with no chance to make a detour or even pull over to the side for a rest, unless you really make it your business to try hard to do that. If you ask me, the novelists have not given us the answers, but they've told us that any day an answer might arrive—and that how we spend that day will help shape the answer we're going to find. And how you spend a lot of those days—well, I guess they become 'a critical mass,' as they say in physics or chemistry!"

All rather hazy and imprecise, yet he had come a significant distance from the youth who was "whistling Dixie," as he once put it—biding his time until he was "older," when certain decisions would somehow be made. Now he was putting himself in Binx's shoes, wondering as he drove *his* sports car (in Cambridge, not New Orleans) what he ought do with his life, and remembering, as Tolstoy urged his readers to consider, that life can end, and abruptly, even for a young person. This student wasn't anxious to go to graduate school, but he wasn't thinking of "taking a year off," either. The very phrase bothered him. "The absurdity of it," he demurred one afternoon, as if one ever could spend a year of one's time in such a way that it wasn't utterly significant, maybe more so than the years that preceded it or followed it: "That phrase 'a year off' sounds as if you're making a jelly sandwich—after one main course and before another!" Then he turned on me, sitting there, spending time known as "office hours." Playfully, he wasn't going to settle for the convenience of that abstraction, either: "Doesn't this time count too, for both of us? You never know where any conversation will take you." Further, he let me know that the "you" could mean me, hence the possibility that my "office hours" might really be part of my "living hours." I could only agree—feel we both at that moment were the beneficiaries of an existential literary tradition.

·4·

Interlude:

BRINGING POEMS TO MEDICAL SCHOOL TEACHING

EARLY IN 1974 my family and I returned to New England after living in New Mexico for several years, working with Indian and Spanish-speaking children of the Southwest. Shortly after we settled in, I received a call from a friend, the poet L. E. Sissman. He wanted to meet and have a talk. I could tell by his voice that this was no casual request. I was about to suggest a place and time when he made it clear to me that he wanted to see me in my study at home, and sooner rather than later.

The next day, when we met, I heard this reserved, soft-spoken man in his forties beg my forgiveness, initially, for the "self-indulgence" to come—the talk about himself. I was struck by the memorable quaintness of this formality. It provided such a contrast with the assumption so many of us have (and not only with respect to the doctors we visit) that the subject of ourselves is an ideal one for the ears of others. Ed Sissman was polite, considerate, and eminently interested in other people—interested in their ways of speaking, as poets tend to be, but also in their habits, preoccupations, values. He seemed out of sorts that day as he sat opposite me, slipping into a patient's posture—yet he wanted leave to concentrate on what was happening in his life, and to do so in the presence of a psychiatrist, like many another citizen of late-twentieth-century bourgeois America.

He was dying, he reminded me. We all are, I observed with the genuine weariness of someone who had just battled a viral pneumonia—though I was also aware of my patronizing, awkward, foolhardy effort at reassurance. Ed received my statement without apparent animus. He smiled. He nodded, even. But the silence that fell was, I realized, either a measure of second thoughts on his part ("Can I trust this guy to get it?") or a gesture of tactfulness: best to let an interlude separate us from what is to follow.

In a second or two he began again, telling me in a fluent narrative about his recent life as a businessman (he worked for a Boston advertising agency), as a writer (he'd published several books of poems, and book reviews for *The New Yorker*), and as a husband (he and his wife lived in a nearby town). When he'd apparently finished, he looked right at me, his eyes meeting mine. As if to head off any comment I might feel compelled to make, he added another, terser remark: "I also have a life as a patient, and it gets busier and busier these days." I knew then why we were there.

We got together weekly from then on. We met in my study at home. Sometimes, after our talk, we went for a drive, for an ice cream or a cup of coffee. Ed had been having more and more medical troubles with a case of Hodgkin's disease that was not (as some are today) curable. He was also beginning to have trouble writing poems. Lines would come to him, especially in the morning, upon awakening, but he couldn't seem to pull them together into a poem. Sometimes he had no "energy or desire" even to write down the words, the images, the metaphors. Yet he had energy for enjoying his sports car, for good food, for a laugh with friends. He had desire, too, at home, on the streets, or at work: an eye for the beautiful in people, places, things. His prose was as pungent as ever, and as shrewdly intelligent. His monthly columns for *The Atlantic*, titled "Innocent Bystander," were a valuable part of that magazine's mid-1970s life. Why, then, the end of making poems?

I'll have to admit to the temptation to say something that

would sound clever. Once, taking absolutely precise aim at me, my quandary, and himself as the provocateur, he wryly observed: "You're going to use words now, I guess, to explain my inability with them." However, I had not one word to say at that point. I sat nervously, sadly, wondering how to get us out of my office. Suggest a car ride, an ice cream cone together, a walk near a stretch of woods to look at birds and plants? All those actions had silenced talk before. He did not prod me on this occasion. He took charge of me and himself both, saying: "The muse has departed the body—looking back upon its quickening decline." A quotation, I thought, maybe from Latin. His smile assured me he was not that serious. I think he had guessed by then that he had an anxious doctor on his hands, one who needed to receive rather than give guidance—as is so often true, some wise older clinicians had told us in medical school, with the dying and their doctors. As Yale Kneeland, a wonderful physician and teacher, put it: "Wait for your cues, and give them your full attention." Someone approaching death can help the doctor to a reconciliation with the inevitable, as opposed to that fury of distracted busy-work that can mask an attending physician's despair as he or she sees the losing battle to be nearing its end.

After Ed's smile came a shift. He seemed to want to have done with himself as our subject matter of the day, or even to help out a doctor who looked frustrated and forlorn as he wondered what to say and how to say it. He asked me, "Why don't you bring some poems to those medical students of yours?" I had felt useless, thinking that there was nothing to do in the face of what was happening to him. He responded with an agenda.

Ed knew I was just starting a course with the title "Literature and Medicine" at Harvard Medical School; he helped me construct a reading list—novels and stories by sometime physicians, such as Williams, Chekhov, and Percy, and by others: Tolstoy, Eliot, O'Connor. I knew of poems, lots of them, that described the subjectivity of madness, a lesser number devoted to the evocation of illness or to a description of a hospital scene, some odes to doctors, and a few lines that rightly condemned doctors for various sins of omission or commission. I knew, too, that over the centuries poets who fell ill were prompted by their experience to look not only inward but also backward and forward—to ask the most important and searching questions about life's meaning.

Sissman had already seriously addressed those moral and existential questions in his poem "Dying: An Introduction" and in other poems, too—"Negatives," "Homage to Clotho: A Hospital Suite," and "Cancer: A Dream." His "writer's block," as some had called it, came late in his illness. He had already done his inquiry with words, and was still working at it as we lapped our coffee ice cream or drove through the countryside, smiling at sights and listening to jazz on the car radio. His title for his posthumously collected poems, *Hello Darkness,* conveys the calm amiability of this tall, dignified man whose sheer brilliance and erudition had been unnerved by an illness; yet illness gave him cause for powerful, terribly unnerving lyrics, ones that medical students today, over a decade after his death, still find "exactly to the point," as I heard it put in class recently. Ed Sissman's "hello" to "darkness" was quietly thoughtful. He was waiting to see. He was watching for signs. He was coming to no conclusions, leaving a few doors wide open. He called me shortly before he died to tell me he'd not be keeping our next "appointment." He had kept using that word, in spite of my efforts to rid both of us of it. I hastened to offer an alternate date. "No need for that," he told me. Stupidly, fearfully unaware (not only patients "deny"), I pressed the matter, so that he had to end it, finally, by saying a firm good-bye: "I hope I'll see you anon." That last word succeeded—by its slight awkwardness, its antiquity, its rendering of the present and the future, its capacity to break through conventional, temporal statements—in conveying what he had long known better than I: that death had no intention of waiting very long for him. That word "anon" was the next to last one I heard from Ed. I promised, too wordily, an imminent visit, and he said, "Good-bye," and meant it. Not long afterward I was sitting in a church, listening to his friends sing of him, and sitting in a class, talking with future doctors about the singing he did during his stay on earth.

. . .

Each year I teach those medical students Ed helped me prepare for—men and women full of decency and earnest goodness. We read prose, mostly, but every once in a while I tuck in a poem; and with those who get serious about writers and writing (a few do every year) the poems become a mainstay. Many of the poets are well known, even obvious: Robert Lowell and Sylvia Plath telling us of madness; William Carlos Williams speaking of his various patients; Peter Davison (Sissman's friend and editor) writing of alarm and grief as the shadows fall upon the victims of approaching death; W. H. Auden uttering one of his requiems, reminders that the end of a life can move us back reflectively toward the splendors of its middle. Poems that tell of pain or a fatal illness, of melancholy or craziness, of losing a wife, a husband, a child, or a patient, with all the attendant sorrow and bitterness and anger, amount to a substantial collection. I have drawn upon the poems of doctor-poets (Williams, of course, and Dannie Abse and John Stone and Jon Mukand), as well as those of patient-poets who have experienced moments of worry or apprehension, things slipping badly with no hopeful end in sight, or things temporarily awry enough to jog the mind loose of words—Philip Levine's "The Doctor of Starlight," Theodore Roethke's "Infirmity," Karl Shapiro's "The Leg," Richard Eberhart's "The Cancer Cells." But most often I hear Ed Sissman's voice in me, his words, the "invitation to the dance" he had to accept when he went to the doctor, told his story, submitted to the poking, the peering, the gadgetry, the trays of instruments:

> Like hummingbirds, syringes tap
> The novocaine and sting my thigh
> To sleep, the sword play begins.
> The stainless modern knife digs in—

Meticulous trencherman—and twangs
A tendon faintly. Coward, I groan.

He did more than groan. He gave us not only ways to take a finite life's last act, but also words for those high moments of consciousness that philosophers say make us what we are, creatures of language who find phrases that display the I clinging to itself and who know that our turn, too, will come.

Today. Tonight. Through my
Invisible new veil
Of finity, I see
November's world—
Low scud, slick street, three giggling girls—
As, oddly, not as sombre
As December,
But as green
As anything:
As spring.

Each year, when I bring Ed's music, the "spring" of his "Dying: An Introduction," to medical students readying for their own surgical "sword play," I try to share him with these fighters-in-training. I tell them of his mind's liveliness and show them a slide or two of paintings of his beloved Edward Hopper. (He wrote about Hopper in "American Light: A Hopper Retrospective.") Often we ended our times together looking silently at Hopper's stillness: "a shaft of morning sun / Peoples the vacuum with American light." His poems, as he described their birth, came with pictures, early in the morning, out of dreams. ("Don't 'free associate,' " my analyst would sometimes say, feeling derisive toward the wordy banality of my posture and his. "Try to describe what you saw last night before you woke up and remembered to cover it with words.") Like the poet Jorie Graham, Ed saw "erosion"; and like her, he thought a lot about life's meaning. I think he'd admire her "At the Long Island Jewish Geriatric Home," as my students do:

This is the sugar
you're stealing
from the nurses, filling
your pillow
with something
for nothing,
filling my pockets
till I'm some kind
of sandman
you can still send away. As for
dreams,
your head rustling in
white ash,
who needs them?

By dying when he did, Sissman was spared such human erosion. When I read that poem of Jorie Graham's, I realized why Ed stopped writing verse a couple of years before he died. He'd "had a few stark nightmares recently," he once told me, but they'd eluded him on waking. I think they had told him that it was best to stop fighting for such memories, and let others step forward—those whose unblinking vision by day didn't have to reckon with the immediate reality of a night with nightmares. Meanwhile, "today, tonight," we doctors can be grateful to him, to Jorie Graham, to lots of poets who help us stop and listen, hear ourselves walking toward those ward beds, toward our own final bed.

Meanwhile, too, my medical students try to make sense of their stage of that journey, make sense of the long struggle they will soon have with death. Already, in school, they know the burdens and obstacles: so many facts to master; so various the professional possibilities to consider; so many diseases yet to be understood. Time was already their adversary in the premedical years and will never let go of their lives, given all the things to know and do, the tests that follow tests, and even in middle age, when every possible certificate covers every available office wall, the test that any disease can offer the person who takes it on seriously. Time is a prize fought for on the medical battlefield, the doctor trying to obtain it for someone, and death anxious to take it away, the sooner the better. With such a combative future for these apprentices, no wonder a poem can have such a powerful meaning.

For some of my medical students, poems have been a psychological or even spiritual mainstay throughout their education. I have known them as undergraduates, introduced them in my college courses to poems by Howard Nemerov, Elizabeth Bishop, W. S. Merwin, Denise Levertov, and of course Ed Sissman; I continue in my medical school courses the encounter with other poets, even as we meet doctors like Lydgate (in *Middlemarch*) and Dick Diver (in *Tender Is the Night*). In some instances we have continued our informal reading right through the so-called post-graduate period, into the hurry and foreboding of the internship and residency years—taking an hour here and there for discussing a story, a poem. "I carry around that book [a thin anthology of poems] in my pocket," one resident in internal medicine told me. "On most days I have no time to read even one poem. But maybe I'll reach for the book, read a couple of lines and try to remember them. In a week, I'll have a poem under my belt."

James, a resident, had done that with Sissman's "A Death Place." As we sat and had coffee once, he joked with me about the poem: "That opening, those first lines—

> Very few people know where they will die,
> But I do: in a brick-faced hospital . . .

—they are lines written for me, I sometimes think! There are days I never leave the hospital, and I find myself thinking I never will. I'll live here and die here! I'll be walking down one of the corridors—I know them all so well I could be blindfolded, yet get where I need to be—and the lines from 'A Death Place' will come to me again. I'll begin to think I'm crazy, hearing the words over and over, but no, we do that with lines from songs, and those are lines from a 'song,' and I'm singing it along with the poet! I've been living out the doctor's side of that poem, seeing people who are staring at death. I've wished I could share some poems with those patients, but you can't put words in someone else's mouth. Each of my patients finds a way of thinking of what's ahead. Some will read the Bible—there's plenty of poetry there. Others will find a poem in the newspaper, or in a religious card that's been sent them. Or in the words of a relative. A patient of mine was reading Robert Frost. One morning he asked me what I thought Robert Frost was reading when he lay dying. I drew a blank. I wasn't sure how Frost died—suddenly, or from a lingering illness.

"I don't think we give our patients credit enough for the meditation they do when they're very sick. Even when I was a medical student, and was just let loose on the wards, I noticed that some of my patients (people who had no education beyond high school, and a few not even that much) would keep repeating the lyrics of a song, or they would repeat a phrase they'd heard in church—holding on for dear life to some words. That's why poems have meant so much to me in medical school and since then. Like patients, poets are probably holding on for dear life to some words!"

James stopped abruptly and looked out a dusty window toward the cloudy sky. A shrug and a sigh reminded both of us how tiring his work can be; but a faint smile on his face asserted something else—his realization that he'd heard something significant, maybe even memorable, come out of his mouth. Earlier that day James had told me how dissatisfied he was with his responses to patients, how inadequate his words sounded as he recalled them during the few minutes of respite he secured for himself. Now, he felt as if a gift had been given him: "There's a terrible poetry to suffering. Your friend Ed Sissman experienced it and gave it to us in his poems; and I think some of us doctors experience it, too, and we try to say what we're experiencing, and sometimes we can't find the words, and other times we're lucky we run into a poet who gives us the words!"

Medical students keep learning to concentrate, to get to the very heart of this or that matter. They keep struggling to tame life itself, in its excesses, its madness. And they keep being stopped in their tracks by moments of tragedy or great bad luck. No wonder, as some of them tell me, their minds ache to give sharp, pointed expression to what they have seen and heard and felt. Poets try to sharpen the sight, to nurture language carefully in the hope of calling upon it for an understanding of what is happening. A young doctor can hear a couple of lines and know what Ed Sissman and all other patients in each "brick-faced hospital" are trying both to remember and to forget. Poets give us images and metaphors and offer the epiphanies doctors and patients alike crave, even if it is in the silent form of a slant of late afternoon light. As James said: "I was standing in the corridor outside a patient's room. I knew she would be dead soon. The blind was down. It was so damn dark. I looked at my watch. It was four-fifteen, a mid-January four-fifteen. Suddenly I saw a line of light—the sun peeking in from the side of the window, sneaking along the floor, climbing up the side of the bed, falling into a patient's hand. I stood there gazing: I was in a dream. Then I heard the patient move a little. I looked at her. She was looking at the light on her hand. She was moving her hand into the light, out of it. She was smiling a little, playing with light, before the darkness took her. I thought of Sissman; I thought of my sister [who died of leukemia]. A doctor is someone who knows lots of moments like that—if we'd only let them happen!"

Perhaps one way of understanding poets is to think of them as akin to that doctor, observing with wonder as a patient caught light, had fun with it, mourned its disappearance. I think L. E. Sissman knew how akin his sensibility was to that young doctor's—hence his desire to record the hospital scenes he had joined as a protagonist, yes, but also as a willful, watching presence eager to bring light into the shadows. When a medical student like James tells me that an entire course called "Medical Humanities" comes down to a line or two remembered amid the rush and chaos of hospital duties, I imagine that young doctor, poised at a room's entrance, watching a patient make light of light, and thinking of Ed Sissman and his poetry—a poet become a doctor's light.

·5·

Vocational Choices and Hazards

We were a group of medical students who met once a week to discuss a short story or a poem we'd all read. The meetings were a break from the sheer drudgery of the first and second years of endless facts, memorized and all too often forgotten so that new information could in turn be (at least for a time) absorbed. We decided one spring to ask a physician who lived not far away to come give a talk at our school. We thought we ought to check with the dean's office, perhaps get the dean to issue an invitation on behalf of the entire medical school. We went to see the dean; he listened to our request and asked how "the second year" was going. We told him we weren't exactly traveling on easy street. He asked what was giving us trouble. The more we talked, the less he liked hearing us. We should stop complaining, he told us, and get on with the job of finishing up the year. Once we got to the third year, things would change, would be more to our liking. We'd see patients; we'd be learning through doing, not just cramming in pathology or pharmacology.

We nodded as he talked, half agreeing, half cowed by authority. After a while, though, we came back to our starting point and purpose. How about an invitation to William Carlos Williams, M.D.—that he come and read some of his poems, perhaps, or some of the stories he wrote about his medical practice? The dean told us he would be glad to oblige, though he also reminded us that we had just told him how little time we had for anything other than our classes, our laboratory work, our evening studying. Perhaps we might think of delaying the invitation? When *would* be a suitable time? one of us asked, a bit testily. The point was to learn from an interesting and gifted writing doctor, even as we were learning from other doctors in the medical center. Finally the dean summoned his secretary and dictated to her the briefest and coldest of invitations. Was there anything else he might do for us? The question was meant to get a rise out of us, literally, and it succeeded: we got up, said hasty good-byes, and left with our heads angrily lowered.

We wrote to Dr. Williams, though, and he telephoned that he'd be glad for us to visit him, or he'd come and visit our group, but he felt uncomfortable about coming on a formal visit to our "big shot medical center." He didn't think he had much to say to us. "To be honest, I'm scared." Why? "I'm a G.P. out here in Paterson and I don't know a thing about academic medicine. Hell, I never went to college, and I squeaked through medical school, and I've been on the firing line ever since—up and down those stairs making house calls." When we tried to remind him that it was his writing life that interested us, he was puzzled. Would anyone in a "fancy medical school" be interested in "those poems"? *We* were, he heard us say, apparently with as much emphasis as we could muster. But he was "scared" on that count, too: "Doctors have their job to do, and I haven't noticed them taking out time to read poems."

Later that year (1953), we went to see him, and sat with some beers and talked about medicine and the humanities—about what a poem or a short story or a novel might mean for someone in medical training. By then Dr. Williams was seventy and ailing; but his mind was spry, and he was quite willing to share with us his thoughts about his two jobs, doctoring and writing. He sat there, relaxed, recalling some of the more strenuous challenges he had faced during the several decades of his practice among the poor of industrial northern New Jersey, mostly immigrant families newly arrived in America. At one point he reminded us that an important part of our lives would be spent "listening to people tell you their stories"; and in return, "they will want to hear *your* story of what *their* story means." Williams also emphasized the significance of writing notes to himself about his patients.

A few months later, tape recorder in hand, I would ask him to talk once more on that subject. "I don't know what I'd do without those patients! Everyone thinks doctors are good people because they help other people who are sick. But if you ask me, the people who are sick are helping us all the time—if we'll let them help us. How many times I've gotten up and felt lousy; I've felt lousy driving over there, and then I'll knock on the door, and someone opens it, and it's a mother or a father, and they want me to go right to their kid, or they have 'pains' themselves, and you know what, the next thing with me is that I've forgotten myself—isn't that an achievement!—because I'm all tied up with someone else. A kid is telling me what happened, and where it hurts, and what he does to make the pain better, or what he's tried to do. Some of those kids, they're playwrights, they're storytellers! They'll set the scene for you. They'll introduce other people, not just themselves; I mean, they'll mimic people, or try to use their kind of words. They'll say, 'And then he said . . . and then she said . . . and then I said . . .' They'll work all that into their own list of complaints: 'And then, while all that was going on, I felt a little sick; but an hour later, when the house was quiet, I heard so-and-so say something, and that was when I got a real jab of pain, and it was like a knife, pushing itself from the skin right through and through to the bone, that's right, the bone. I could see stars, that's how hard the pain was. I thought God was trying to get me to kneel and plead with Him to come get me, and I mean right away.'

"Do you see what I'm trying to say? I can't hear a kid talk like that and not be sprung—sprung right out of my own damn self-preoccupations. I'll pick up a good story or novel and the same thing happens: I'm in someone else's world, thank God. I'm listening to their words. My own words become responses to what they say—the novelist or one of my patients. We're on stage! I don't mean to demean medical practice, or turn it into a dramatic production. I'm thinking of the moral seriousness [you can see] on the stage, a certain kind of exchange between people, where the words really are charged. Sometimes when I'm with a patient who is having trouble getting across to me what he wants to say, I tell him to stop describing the pain, and just tell me where he was when the pain came on, and what he was doing. I say to him: When the pain knocked on your door, interrupting your life, what were you doing? I try to get them to talk about that life before the interruption, and then as they describe the interruption, I can get a better picture of what happened than if they spend their time trying to find the right words for the bellyache or the chest pain.

"When I try to get them to take medicine or change their habits, I use my own life for a while; I tell them about my medical problems, or my wife's, or I tell them about others I've treated, their struggles. I don't (I hope) run off at the mouth; and with some I can be almost dead silent. Some are comfortable being stoic and mute. A prescription is all they want! But most of my patients—they want to gab away, but they're not sure how to get going. They're in trouble; and that's when you're eager to look into things deep, real deep. I wouldn't walk away from those kinds of talks for anything; I come away from them so damn stirred myself—I've needed to walk around the block once or twice to settle down, or drive out of the way for a block or two, so I can stop and think. It's like reading Tolstoy; you can't just breeze through his stories, even if his writing is so 'easy,' it just envelops you. You get stopped in your tracks by something he says, and it takes time to let it work its way through your head and your heart, both of them!"

He paused, as if certain memories were still there, on the top of his mind, in the very midst of his heart. He resumed talking —now, about medicine as a profession, its risks as well as its opportunities. "Too many people see us as big shots," he began. He laughed at this slangy choice of words, not a rare choice for him. The phrase "big shots," as used by him with a touch of sarcasm, seemed to serve almost a medicinal purpose.

In his poetry and short stories he introduces a doctor who is himself: a busy, burdened, vigorous person, able and effective one moment, vulnerable the next. He wants the reader to know this doctor, warts and all—his irritable outbursts, his petulant and peremptory moments, as well as his conscientious, caring side. He wants to warn himself, or the reader, that failure in medicine is not only a matter of illnesses unsuccessfully treated: "I can recall times when I made the right diagnosis, wrote the right prescription, and was awful with the patient. I was in a lousy mood, and the poor patient had to put up with it. I'll never forget the old lady—she was Italian and spoke only broken English. She watched me with her daughter and her granddaughter, talking with them, and she waited for me to put my stethoscope back in my black bag and get ready to leave. She came closer to me—she'd been in the back of the room—and she told me she was going to pray for me. I looked at her as if she was a little nuts! I thanked her, but she could tell I wasn't understanding her reasoning, and she damn well wanted me to understand. So, she explained herself. She told me I was in a hurry, she could see, and I was speaking 'rough'; that was the word that got to me. She was an illiterate old lady, but she knew the right word, the exact one: she had me picked out, pinned against the wall. I threw up my hands. I surrendered. I gave a big laugh, the kind that said: OK, thanks, and I'll try to explain myself—but first listen to my nervous noise-making while I prepare my speech!

"Then I told her, in the simplest words I knew, how busy I was, and worried about my sons (both of them were sick with bad colds, and one was getting pneumonia, I thought). When the daughter thought her mother missed something I said, she translated it into Italian. When I was all through, the old lady gave me a smile. She was wonderful; she became *demure.* That was the word that came into my mind back then. I can remember thinking of the word as I looked at her. She had a mixture of shyness and real toughness in her. She was going to teach me a thing or two! But she could see I was no total dope. She could see I knew how to go about my business. She just wanted me to realize that everyone isn't going to be a busy doctor's slave or his patsy or his complete fan. She was trying to be *nice* to me, I suddenly realized. She cared enough to leave the shadows of that room, to come right into the ring and throw a few jabs.

"We became friends. She sent me home, a few weeks later, with her pastries and her wine. Boy, did I watch my step with her family—and with all the neighbors in that building. If I felt grouchy, I'd say to myself: Be careful, or you'll get caught and put in your place around here! It was easier to hide behind my professional prerogatives when I was in Rutherford, with folks like myself, the middle class: with us, if a doctor is silent, even a touch rude, it's just his 'style,' and hell, all that matters is whether he can cure you or not. We put up with the biggest jerks, make excuses for them. I guess we're really making excuses for ourselves—why we let someone treat us like that, humiliate himself and us both. That Italian lady would have none of it, a doctor's arrogance!"

Such an approach to professional matters, such a confessional mode of self-analysis, self-presentation, was vintage Williams. I heard him many times confronting his flaws, and I read his efforts to render that aspect of his life insofar as it informed his medical practice, in a series of stories, first published in 1937 as *Life Along the Passaic River.* When I began teaching college students and medical students, I found those stories powerful in their impact upon those young men and women who, like Williams, struggled with ambitious intelligence as a force that can demolish the "heart's reasons"—namely, a warm empathy, a considerateness toward others, a willingness, even, to let them become one's teachers, however humble or troubled their lives.

In Williams' stories the reader meets a New Jersey doc, informal and observant and often harassed by the demanding obligations of the practice of family medicine, including house calls. What we learn is not only the nature of Williams' practice—the constant challenge of diseases as they occurred in the 1930s, before the arrival of antibiotics and other breakthroughs—but the nature of the doctor's response to that kind of professional life. This doctor, in story after story, admits to moments (and longer) of pettiness, impatience, annoyance, anger, outrage, disgust, prejudice; and also to times when he is frightened, anxious, or, yes, excited, even aroused, by a patient's appearance, or manner of speaking, of standing, of responding to him.

Such attitudes are hardly the stuff of everyday discussion among medical school students and teachers. One of my students, reading the pieces that a few years ago were collected, arranged, and published as *The Doctor Stories,* referred to the subject matter as "the great unmentionables of everyone's everyday life." I suspect Dr. Williams would have smiled at this description—for that was precisely what he meant to write about. All through his greatest work, the long poem *Paterson,* he poses one of the oldest dilemmas: the sad commonplace that ideas or ideals, however erudite and valuable, are by no means synonymous with behavior, that high-sounding abstractions do not necessarily translate into decent or commendable conduct. People can, as he put it, "talk a big line" and "come out badly wanting in their actions."

Once, discussing medical school education, his own and that of students learning to be doctors in the 1950s, he made clear the direction of his thoughts with respect to the subject of "medical ethics" by recalling some personal memories: "You can set rules; you can teach lessons; you can give tests; you can pass them, even pass with flying colors—but even so, stubborn human nature is out there, threatening to take charge of the intellect. I wish some of us old docs were let loose among you medical students. Hell, I wish we knew how to live with one another, never mind you folks. There's a big difference between our high talk, though, and how we behave ourselves when we're out there on our own—and we should make that the italicized preface to every lecture, every piece of advice we hand out. It's too damn easy to teach, to preach, then go off and be your own, full-of-yourself self. I speak with the voice of experience! Maybe I'm too hard on myself—a little. But who can take for granted his good days? Who can honestly say he doesn't stumble more than 'now and then,' especially when there's no one keeping score?

"I'm sure most of us docs work hard and try to do the best we can. But I'm not sure we don't hurt a lot of people with our manners, our sour moods, or the big rush we're in. I don't have answers. I know we've got a lot on our minds. I don't need someone reminding me how tough this work is; I *know* how tough it is, from years and years of experience. But—a big but—you're on thin ice when you hear yourself apologizing for yourself and begging for the next guy's pity: See me, how hard I work, and the good I do, so shut up if you have any bellyaching complaints. That's the most tempting line for me, the one I take when I'm in trouble. It works like a miracle. People leave you alone! They feel guilty, or they get worried: someone's life could be at stake, and I'm bothering a *doctor,* this doctor, with my preachy message that he should be kinder and more sensitive to me and my relatives!

"Of course, if anyone gets too bothersome, I can always put it to them: Do you want someone who is full of honey and doesn't have the toughness a doctor needs, or do you want me, vinegar and all, doing the hard job I have to do? Now, how's that for a choice? I hear young doctors in the hospitals all the time talking like that to each other—they're really talking like that to themselves. They're saying that good guys don't win races, that a doctor is a very busy fellow and he does hard work, and he can't get 'soft' or he'll fall into a kind of incompetent sentimentality. Even pediatricians and psychiatrists talk like that—how you have to be 'cool' and keep your 'distance,' and not become 'overwhelmed' by all the emotions that come your way. Of course, it's a clever setup: who in hell is suggesting that anyone should be overwhelmed, or should get soft in the head or self-indulgent emotionally? Why should we always be told that the alternative is between a doctor who really knows what he's doing, even if he doesn't have much time to be with his patients, to talk with them and be understanding of them, and a doctor who has all the time in the world for his patients, but he's a first-class idiot, and could end up being a threat to your favorite relative's health, even life? I'll answer that. It's not a question; it's a rhetorical statement meant to rationalize callousness and egotism!

"For crying out loud: who in hell wants a dope hanging around with a stethoscope? But why is this dope conjured up every time some swaggering tyrant or mean, cold son-of-a-bitch, who happens to have an M.D. to his name, shows up and starts bullying people? I guess it's because the rest of us [doctors] don't want to see him get what he deserves. And why? Hell, we know our own dark side! We rally around others to protect ourselves! A pity—because lots of us, I'll say most of us, don't need that kind of protection. Who are we, a bunch of moral monsters? Maybe we're tempted to be, sometimes. When people are scared as hell, and death's around the next corner, or maybe facing them, right in front of them, they'll grab at everything there is, and we're what there is, what's available. We do battle with the Devil himself, if you want to trespass onto theological territory. So, we're gods for others—but we know how tinny we can be, or we damn well should know. I guess the raw truth is that the worst of us don't know: the ones who strut and prance and con themselves and everyone else into thinking they are God's hand-picked emissary, if not chosen successor.

"Sure, if you could get medical students to read certain novels or short stories it might make a difference. But, I'll tell you, don't bet too much of your money on it, because you know what can happen with any book, even George Eliot's *Middlemarch,* or Chekhov's stories. Have you noticed what goes on when literature professors get together in a room? Are they saved by George Eliot or Chekhov, by Shakespeare or Dickens or Hardy, or even by Dostoevsky and Tolstoy, who tried like hell to save us all, poor bastards that we are? Tolstoy had one hell of a time trying to live up to his own ideals—and boy, did his wife and kids have to pay for the moral struggle he waged. There must have been plenty of times when he was worse than the most arrogant doc you and I can conjure up in our imagination! So it isn't 'the humanities,' or something called 'fiction' (or 'poetry') that will save you medical students or us docs. Books shouldn't be given *that* job, to save people, to improve their psychology, or their manners, or the way they talk with their patients."

. . .

At that point we were interrupted by a phone call. One of Dr. Williams' patients had suffered a relapse: a young woman of fifteen, on the road to recovery from severe pneumonia, had spiked a high fever yet again. She had a past history of rheumatic heart disease, with a consequent heart murmur and the long-term prospect of congestive heart disease. All of that and more Dr. Williams explained to me as he hurriedly drove to her home, a tenement building in Paterson, where poor families struggled hard to get by—make enough money to pay the rent, buy food, stay warm, and, yes, give their doctor, who made house calls, an occasional payment for his services. As we climbed those dark stairs to a fourth-floor apartment, the doctor suddenly whirled around and looked right at me, his eyes meeting mine. A moment's silence, and then he said: "I don't like this mother you'll see. She drives me crazy with her questions. I want to take the girl and save her. I want to take her younger brother and save him. I don't like the father, either. What a pair! The miracle is that wonderful daughter of theirs. She puts up with them, and she sees everything, all their bad habits."

A second later, he was rapping his right hand's knuckles on a door whose brown paint was peeling and whose splinters were noticeable—he had to be careful about where to make contact with the wood. Three knocks, then a shuffling of his feet to register impatience. Noise arose on the other side of the door, and I could see his face suddenly concentrating, the ears perking up, the head unselfconsciously tilting toward the door, as if he was interested but didn't want to admit to himself, never mind anyone else, that he was *that* interested—and so, abruptly, he moved away a step from the door just as it opened. (I later realized he'd probably heard, among other things, the sound of approaching steps.)

The mother opened the door and, without saying a word,

stepped aside for us. I can still remember her eyes meeting those of Dr. Williams. My heart went out to her; she seemed retiring, frightened, and eager to please. She motioned to her daughter, asking her to get ready for the doctor by sitting up, and then moved away to make room for him. But the girl, febrile, seemed uninterested in what the mother wished. Meanwhile, Dr. Williams put his black bag down beside the patient's bed, glanced around the room, caught sight of a chair, pulled it toward the bed, sat down, and took the girl's left wrist into his right hand and held it firmly. At that point she picked herself up and sat half upright against the pillow, staring at her wrist and Dr. Williams' hand. He asked her how she was doing. She was doing OK, she said. He wouldn't accept her answer: "Tell me the truth." She immediately changed her reply: "Not so OK." He nodded, let go of her wrist, told her he was sure, from her pulse, that she had a fever, and announced he would soon examine her. He explained that I was a medical student and asked if she minded my being there as he listened to her chest and heart. She shook her head and smiled. It was then, as he reached for his stethoscope, that the mother entered the scene, so to speak. She drew nearer, frowning, and addressed the doctor: "Why no temperature?" Suddenly I realized I hadn't heard her voice before, and realized, a second later, that her English was imperfect. She wanted the doctor to take her daughter's temperature, to learn whether she had a fever. He barked back at her: "Leave her alone!"

I was stunned. He spoke with obvious annoyance. I knew he could be cranky; he had told me so, and I'd seen him moody and irascible at times, especially when we talked about certain literary matters. But this situation was puzzling, and for a moment I wondered whether he wasn't going beyond the ethical bounds, being rude to a patient's mother for no justifiable reason. Then came her retort: "You not do right. You always do wrong thing. She hot. Give her the thermom." While my ears concentrated on that abbreviated last word, the doctor pointedly turned his back and started doing his examination with the stethoscope. It was left for my unobstructed and idle ears to hear a tirade from the mother, uttered in a mix of broken English and fluent Italian. She had come close, very close; by the time he had finished his examination, she was nearer the doctor than I. As he looked up, her eyes were ready for him, for his eyes. Again they locked into a mutual gaze. It was then that the patient looked at her mother and said her own imploring word: "Please!" The mother backed off, and the doctor began telling the girl what he believed was happening—the pneumonia was returning. He urged fluids as well as aspirin. He asked whether she'd been taking the medicine he'd prescribed earlier. No, she had stopped. Why? The answer: "Because my fever went down, and she [the mother] told me I must stop taking the pills, since I'm cool."

Now the doctor understood all—and became enraged. He turned to the mother and opened his mouth, ready (I thought) to give her the lecture of her life. I could see the color on his face. I could also see his right hand tighten its grip on the stethoscope. No words came out of his mouth. He simply stood there, bearing down with those formidable eyes on the woman of forty or so; she didn't budge, however, nor did she stop looking right at him. I wondered who would flinch first. Neither did. He suddenly flung up his right arm, stethoscope in hand. All eyes in the room, the patient's and mine as well as the mother's, followed the course of that ancient medical instrument, that symbol of a profession's authority. With that gesture the mother retreated, and then the patient slumped back onto her pillow. Dr. Williams started getting ready to leave. He folded his stethoscope carefully and put it into his black bag. Only at that moment did I begin to relax a bit—the weapon was being laid down. He turned his back on the mother yet again and began a quiet appeal to the patient—she ought to take those pills every day, as he had instructed, and not stop taking them under any circumstances whatever. He emphasized the importance of that last remark with his voice, and after he'd finished speaking, he looked directly at the girl and asked her if she understood what he had said. Yes, she did. Then he asked her why she hadn't followed his instructions and kept taking the pills. Silence. He lowered his head. But he wasn't feeling shy or humble; he was fuming. Finally the head was raised, the mouth opened, the body turned toward the older woman: "I want every one of these pills taken by her. Every one." He was pointing at the patient. His eyes once more engaged with those of the mother. Neither retreated until the daughter addressed the mother: "Please pay the doctor." It was then that the mother left the room; she came back with a tightly folded bill, which she gave to her daughter, not the doctor. The daughter immediately handed it to the doctor, who quickly put it in his pocket without so much as looking at it. I remember being curious as to the amount. In a few seconds we were out of that apartment, and shortly thereafter on our way back to Williams' Rutherford home.

He didn't speak, and I was afraid to say anything, ask anything. Usually he was a willing, animated talker; now all his energy seemed given to the act of driving his car. He held the wheel with both hands (he often used only one hand to drive), and he fussed with the windshield wipers, putting them on and then turning them off as some rain fell, then stopped falling. Suddenly came his first words: "Hell, make up your mind" —addressed, I realized, to the nameless, faceless weather. Then he laughed, turned to me, relaxed a bit—his back slouched against the back of the seat—and began to talk: "Well, there's a lesson for you in medical psychology, in what a doctor has to put up with." He wasn't satisfied with what he had said, so he went into a long explanation while the windshield wiper worked away at nothing, the rain having stopped. I was told all about the family, the patient's medical history, the mother's psychiatric history, the neighborhood's (immigrant, impoverished) social and cultural history. I was also told about previous encounters of the kind I had just witnessed. The narrator was sometimes wry or even humorous, at other times sardonic and angry. During one of those latter spells, he pulled out the crumpled bill he'd been given, telling me as he did so that it was "only one dollar." He laughed, with a mixture of embarrassment and annoyance, at what he'd heard himself say, and he added: "I'm lucky to get anything at all! Be grateful, I guess. Don't envy your colleagues what *they* make!"

Hearing himself say those words gave him pleasure. The smile stayed on his face as he shot me a car driver's quick glance. Soon he was turning his attention to me, to my younger generation, to the matter of how doctors ought to be educated. He was quite talkative, and I only wish a tape recorder had been there, dangling, perhaps, from the roof of his car. What I remember, though, was the passion, at times the vehemence, of his remarks. "Try not to get *too* carried away with yourself—but you will, time and again." And a block or two later: "I wish I'd been a stronger person back then [in the apartment we'd just visited], but certain people, and certain situations, just get to me—and that was one; *she* was one." After he'd made that confession and apology, he was quite eager to talk in general about the experience of medical school, and specifically about the subject of medical ethics—what he called, interestingly, "the psychology of the professional man."

He made a distinction between himself the doctor and himself the writer. It was easier for him to diagnose his blind spots and prejudices in a story than to "cure" himself, he said. By that he meant "head off" his "sins" before "committing them." He was a bit playful about the use of the word "sins." He reminded me that he wasn't particularly religious—yet at times, he said, he preferred the word "sins" to a psychological equivalent, such as "problems." Why? He believed the issue was often moral—it was a question of the moments of wrongdoing that take place when one person can't stop himself from being inconsiderate of another, or even mean. He noted that even his doctor friends who had been analyzed, and indeed who were psychoanalysts, admitted to failure when they were deaf or blind to a patient. It is important, he insisted (at one point pounding the steering wheel to emphasize the intensity of his feelings), that doctors keep reminding themselves—and be reminded by others—of the power that goes with their profession, and the grave temptations that accompany such power. He talked also of a "moral drift"—

spells when he had forgotten the harm he might have done others, and thereby himself as well.

Later, in his office, he amplified that point at some length: "We have got to remember that when we give the back of our hand to others, we are hurting ourselves. When I spoke of 'moral drift' I meant an indifference to others that can become a habit. So we keep doing it, and we're without a sound moral compass—once, twice, thrice, and soon you stop counting! It's too easy for lots of us to forget what it's like for our patients, to be damn preoccupied with ourselves. I don't want to be too tough on doctors, on myself. This is no easy job we have; but that's no excuse for taking out our frustrations on our patients, and lots of times we do. Hell, *I* do. Never mind this 'we' business!

"It's our 'little moments,' I'm convinced, that will get us at least at heaven's door, or send us straight to the Devil's fires! I swear, some of us can keep being 'correct' in the decisions we make, and yet be tough bastards who treat people with arrogance and condescension. I wish we had medical ethics courses that pushed us to take a hard look at ourselves, not just examine the rights and wrongs of those 'situations' that are often posed in clinical conferences: was it right to stop 'treatment,' or wrong, and why? I think *Arrowsmith* or *Tender Is the Night* are trying to get at something else—the way a doctor's general attitude toward people, his personal decency and his view of what life means, can influence the way he practices medicine. Sure, every once in a while we'll have a big medical puzzle—for example, a patient who is dead, for all practical purposes, but the heart is still going. The next thing you know, all the docs are talking about medical ethics. They want to have a long discussion about the patient; and you'll hear pious rhetoric about 'rights and responsibilities'—the patient's rights, the doctor's responsibilities. It's a great time for a kind of low-level philosophizing, and hell, why not! I've sat in many a medical grand rounds discussion that ends up being just that—doctors sharing their ideas with each other on what this life is all about. But the focus is usually on some relatively rare medical situation, or some technical question: is someone dead when the brain's waves stop, or when the EKG is flat?

"What [F. Scott] Fitzgerald is getting at—oh, the tug of the heart, the madness in the head, a different level of reality than the kind an electroencephalogram or electrocardiograph machine records. Dr. Dick Diver, that forlorn shrink of *Tender Is the Night,* has a lot to tell us doctors about what can ruin us in our work—or as it's put when you pick up a program at a medical meeting, the 'occupational hazards' we face, doctors on the line. How's your marriage going? How are your kids doing? Are you keeping up with the bills? With the Joneses? You laugh! If you and I could be flies on the wall, sit in doctors' offices, and keep track of those same doctors at home, when they're with their wives and sons and daughters, and friends—I'll tell you, there'd be correlations. A bad scene in a marriage, for example, would correlate with a tough scene in an office—the doc berating the patient, or being pretty short-tempered, thin-skinned. Now, let me tell you: it just won't work to pull that fellow aside and tell him to read a novel or a poem! Sure, maybe he should go see a psychiatrist. But there'll never be enough psychiatrists to go around, and these everyday ups and downs aren't meant to be taken to the analyst's couch. Hell, my analyst friends tell me of their ups and downs, and I'm sure their patients know them [the ups and downs], even if not in a completely conscious way!

"What do I recommend? I've got no solutions, only a few obvious ideas—that people should talk to each other in a medical school class the way we're talking now. I'm saying that the more open we are about what gets our moods going, and how those moods affect our work, the more likely we are to catch hold of ourselves—in the nick of time. I don't want to compete with people who go to priests and say *mea culpa* (lots of my patients tell me they stay alive because they can say those words to their confessors!), but there are days when a few bad lines on a typewriter translate into a bad doc. I'll be morose, grouchy, out of sorts, moody, sullen—you pick the word and it'll be the right one to describe a son-of-a-bitch doing his job, all right, but with a mind that's clobbering him, and with no heart at all for people who need heart as much as they need to have someone listening to their hearts. That's why, when you say you want suggestions, I can only come up with my shame, as I remember it, and its sources; and I can only say: let's have some heart-to-heart stories to tell each other, the folks who teach medicine and the folks who are learning it."

Dr. Williams was constantly urging me not to allow natural egotism to obstruct a larger view of what it is that any profession offers in the way of moral possibilities and hazards. So often, he pointedly reminded me, students soon to become doctors, lawyers, architects, businessmen, teachers, or engineers are understandably preoccupied with their performing selves, with matters of technique, of knowledge—even though, he insisted, "it is your response to the ethical questions that will make you what you are." What in the world did he mean by that remark? In those days I assumed that if I knew how to be a reasonably competent physician, I'd not be in any moral danger. True, we all knew as medical students that we ought to be as courteous and compassionate as possible—but such an attitude was, mostly, assumed rather than closely examined or asserted. Indeed, time and again deans and doctors inflated our moral and intellectual credentials. We were "carefully chosen"; we had overcome "many obstacles"; our "ability" was "never in doubt"; our "capacity to become good doctors" (a phrase used over and over in speeches to us) was, as one university official put it, "obvious and a matter of record."

Dr. Williams laughed at that last phrase—at the implications of such an assumption about a bunch of men, mostly (the overwhelming majority of my medical school class were white, Protestant males of comfortable backgrounds), who had little more than two decades of a somewhat cloistered school life to their credit. He scorned, for one thing, the notion that an academic "record" said all that much about anyone's character, his or her moral life as it got lived from day to day. But he went further;

he challenged the use of the word "obvious"—and with gusto, not to mention a certain heated truculence: "Hey, you guys are lucky, and I'm sure you're all well brought up, and I'm sure it shows, as the saying goes; you're a bunch of show horses. You've been tested (I'll say!), and you're in for the long run. You'll last! You took courses you had little use for, memorized everything in sight, did well, promptly forgot what you crammed in, and then were rewarded: you 'got in.' Wonderful! I suppose that *is* 'obvious'; and I suppose it *is* a 'matter of record.' I'm sure each of you had five bouncy, effusive letters of recommendation, written by those folks who tested you, or knew you for other reasons: you worked for the guy, or he's your dad's friend, or his son and you went to school together. But for crying out loud: what do all those grades and letters tell us about how you're likely to behave when you're sitting on a bed, and there's someone else on the bed, and it's not someone you're wanting to sleep with—I *hope*, though Lord knows I won't automatically assume, that it's 'obvious and a matter of record' that you won't want to sleep with some of your patients!

"I wonder what would happen if all you fellows were put to work this way—sent into a neighborhood where there were some needy people and told to go try to be of help. You would tutor. You would clean up the streets. You would go give a hand to someone who's paralyzed, or is too old to get out. There's a lot you would do, if you had the moral energy to go look and keep looking until you'd stumbled on your answer."

Again he posed the matter of "conduct" (as opposed to "ideas"), repeating, as he had in his writing, that the ultimate test of a person's worth as a doctor or teacher or lawyer has to do not only with what he or she knows, but with how he or she behaves with another person, the patient or student or client. With some frustration and subdued irritation I asked whether there was any way to avoid the ironic impasse of teaching a *course* on ethics. After all, the students would try hard for the reasons they always did, and perhaps the most persuasive, eloquent talkers and writers of examinations would end up being cold, surly (and who knows?) mean or malicious doctors, teachers, lawyers. He agreed that such a possibility had to be acknowledged.

He was impassioned, but by his own admission not totally convincing, since, after all, "there are no guarantees in matters of the heart." He was reminding me that we weren't talking only of an intellectual question. We were searching for a reliable thoroughfare, maybe—a direct passage from the world of thinking to that of day-to-day living. If we would find no road for everyone, we might at least prompt a substantial number to make the trip through the kind of "enchantment," as he put it, a novelist such as Dickens offers a willing reader. (Bruno Bettelheim would use the same word for a similar purpose years later.) Perhaps Williams was overly confident that somehow the storyteller can exert a pull on readers strong enough to win them over to a way of getting on with others that is morally inspired. He knew, of course, that readers can be hasty, devouring, unresponsive to a writer's intent. He knew that a gifted novelist can tell a wonderful story, one full of moral power, and yet himself or herself fall flat on the face when it comes to the test of how the personal life measures up to the written word.

Still, as the old doctor kept saying, "we have to keep making the effort"; and "the more palpable the connection between the story and the reader's story, the better the chance that something will happen. Look, these novels or short stories aren't meant to save the world. But a story *can* engage a reader—not every reader, and some readers only somewhat, but plenty of readers a lot, a whole lot. I mean, art reaches the mind and the heart, and in a way that it doesn't easily get shaken off. My stories about my own medical struggles are, really, the stories of my patients as well. They've recited them to their wives and husbands and friends and children, and I've heard them, just as I've recited my version to my wife and children. If I've failed at times as a doctor (and I sure as hell have), please remember that I may also have passed a few times. It's being made uncomfortable by a story, and trying to put it out of your head—a story on paper, or someone's statement, which can be part of someone's story. But you can't quite succeed—banish the damn words—because you've been influenced, and the day may come when that influence wins you over."

. . .

During one of my blinder, dumber spells of teaching—I was holding office hours but was in a rush because of a writing deadline—a student who had sensed the psychological and moral deafness that had taken over my working day remarked to me that I had really taught him something during the past few minutes of our conversation. I had been trying to squelch a yawn, and was also surreptitiously glancing at my watch. But now I was awake and politely asked the young man what it was that he had just learned. I prepared myself guiltily for his gentle reprimand, but he had his own agenda to pursue.

"I was asking myself on the way over," he said, "why I was coming to see you when I haven't really given myself a chance to sit and struggle with this chapter [of *Let Us Now Praise Famous Men*]." Then he added: "Agee warns against falling into conventional ruts, into lazy habits, and here I am, doing just that." I suspected for a moment that this college junior was not in fact indifferent to my apparent indifference, and so I apologized to the student for my not being altogether "there." I offered my reasons. Surely a personal statement, a more or less explicit confession, would make amends. But he remained quietly intent on his own moral introspection, and soon enough I realized that his struggle had indeed been with himself, and had preceded his arrival at my office door. He asked me: "When you graduate from college and read a book, whose office hours do you visit when you have a question?" I was sure he was teasing, but he went on to respond calmly to his own question. He pointed out to me that sometimes teachers elicit dependence rather than independence in their students. He then wondered aloud whether every profession didn't pose such a danger to those in it—doctors becoming all too sure of themselves, or lawyers or architects. True, he acknowledged, those who are in a profession most definitely do possess a body of knowledge, training, experience—and yet the temptation for everyone else is to surrender.

Actually, the whole context of this conversation had a great deal to do with what the student was saying. When he arrived in my office I was so preoccupied that I didn't listen carefully; my sense of guilt over my own attitude made me ascribe to him a hostility and defensiveness he was not expressing. My assumption that I knew what he was thinking was in fact the very kind of paternalistic attitude—one shared by many professionals—that he *was* talking about. We become parochially authoritative. In *Anna Karenina* and *Little Dorrit* it is the bureaucrat, the government functionary, whose coldness or self-importance is depicted. In Eliot's *Middlemarch* and Chekhov's story "Ward Six" and O'Connor's "The Lame Shall Enter First," it is the doctor, the healer, who becomes morally compromised by indifference or egotism. In *Bleak House* and *The Trial,* those who prosecute others are morally blind, and a similar theme runs through *Invisible Man.*

Ibsen's *The Master Builder* develops in great detail a psychology of megalomania, explores how grandiose we can become under the guise of exerting a visionary talent on behalf of others. A graduate student in architecture told me of his reaction to the Ibsen play: "I've felt what he [Ibsen] is getting at when I sit with my designs, my plans, and imagine a whole block, a whole neighborhood, a whole city, 'made over' by me. There's no end to what I can do, *should* do. I'm being negligent if I'm not ambitious—and I mean, as we put it, 'on a big scale.' Think big, build big. Even a small effort, a building others might think of as insignificant, becomes in my imagination an 'enormous challenge.' Why do I keep hearing people use that phrase? What do they mean when they use it? I'll tell you: I am an architect—a 'master builder'!—and I'm not going to get caught 'dreaming low.' That's what one of our professors worries about—that some of us will fall asleep at the board and end up 'dreaming low.' He wants us to 'think of everything'—the way he puts it—when we do our sketches.

"He's right. Architecture is both an art and a science. We have to think of the smallest details—how everything will work, or *whether* it all will work. What will happen, for instance, when glass is used on a room's ceiling—in the winter, with snow piled high? But we're also trying to be original, to be admired as creative. Some of us talk about changing environments, about building cities. Ibsen got it right—the ambitiousness of it all and the dangers that can result. We're not influenced by 'emotions'; we just design. We're like surgeons, I suppose—the big-headed, swaggering surgeon. Everyone knows about that kind of surgeon, but only a few architects are considered to be 'carried away with themselves.' I'm not saying all of us *are*, but I know how tempted we can get to push our clients around, to intimidate them or take advantage of their innocence or gullibility, or, worse than that, to play up to *their* pride. What a vicious circle gets going: the client who is full of the desire to impress the world—and has the money to turn a desire into a reality—and the architect who has the same kind of mind at work!

"Only one teacher here [at the Harvard Graduate School of Design] ever addressed that question, at least in my experience, and it wasn't in a classroom. We were at a party, and he'd had a couple of gin and tonics, and he joked with us—that we should beware of an architect who gets drunk. We thought he was being charming about the gin he'd just swallowed, but he had something else in mind. He poured out a psychological portrait of his mentor, one of the big names in architecture; he told us how visionary the man could be, but also how ruthless and manipulative. It was some moment! I've never forgotten it—though to be honest, nothing I heard was all that new or surprising. From day one here [at graduate school] I've learned to deal with the egos of bright and gifted teachers, and I suspect it's the same at other Harvard graduate schools, or at any institution. Architecture, too, comes down to human beings, in all their variety; it's not just the abstract design, and then the bricks and mortar of making the design come alive. It's power and money and clout and vanity and bullying and conning—some of the time, at least."

The speaker was convinced that he was not only being relentlessly candid, but offering a rare and idiosyncratic vision to his listener: "It's hard for us to level with ourselves, especially for an architect, because we're not the kind of people who do a lot of—I guess introspection is the word." Others, however, in his profession had shared such moments with him, he acknowledged, and indeed his claim of distinctive forthrightness is itself, ironically, a common denominator among those in a number of graduate schools who are having moral qualms about their chosen profession. Again and again I hear the bluntest criticisms made of professional education by soon-to-be doctors, lawyers, business executives, teachers, architects, government bureaucrats, and I hear suggestions for remedial action: courses in ethical reflection; courses in which students become part of a "psychological group" in the hope that "confrontation" will take place. So often, those who are talking as I am, urging what I am urging, are the ones who are already converted. The real problem, they tell me, is to reach "others," meaning the students who aren't interested in questioning their conduct as overbearing, presumptuous, or imperial-minded architects, or as Dickens' grasping and selfish lawyers, or as doctors who have little real respect for patients—doctors of the kind Chekhov portrays in "Ward Six," or Flannery O'Connor in "The Lame Shall Enter First."

• • •

That story of O'Connor's, I have come to realize, inspires a good deal of thought in all sorts of students. Undergraduates enjoy both her seriousness and her humor as she fleshes out the Biblical injunction "Physician, heal thyself"; medical students nervously recognize their own possible future as they contemplate the healer who neglects his son while lavishing inordinate attention on someone he is trying to help. Students studying law or business administration, or those enrolled in a graduate school of education, are also responsive to the story because its central irony has application for so many of us: policemen whose children become lawbreakers, cobblers whose children go shoeless while they hurry along repairs for customers, teachers whose sons and daughters receive inadequate educations. O'Connor's therapist chooses a bright patient to favor with his attention. Before the author is through with that therapist, the moral import of the story is clear: a professional person's vanity is a critical aspect of his life. His desire to give so much of himself to a particular patient has to do with his sense of his own importance. Others might not be worthy of such a patient! As one of my medical students reminded me during a tutorial discussion: "I hear doctors brag about their patients the way I used to hear teachers brag about their students—and the way I have heard my father [a lawyer] brag about certain clients of his for years." Another student in the tutorial group dared make an unforgettably terse formulation: "Someone becomes a mirror to a professional person."

In "Good Country People," Flannery O'Connor chides herself; in "The Artificial Nigger," she chides her readers. Those two stories also lend themselves to teaching in any part of a university because they address the constant temptations toward egotism and self-deception to which the mind is heir. In both stories the person who is supposedly smart turns out to be morally vulnerable. Hulga's self-importance in "Good Country People" hints at a personal acknowledgment on the part of the author: Hulga is an intellectual woman in a rural setting who risks becoming too clever for her own good. Mr. Head in "The Artificial Nigger" is anyone of the intelligentsia who is sure of what he knows but may well be proven flawed by an ordinary event on an ordinary day. His grandson gets his education, all right: he learns that one can be sold short even by close kin, heady though they are.

A student of mine, a year short of becoming a physician, told our seminar that he could imagine himself reading the story "The Artificial Nigger" every year until he died, and each time being reminded of "the same truth." When we asked what that truth was, he replied with memorable words: "That we're always in danger of talking one line and living another—and the more successful we are, the more the danger." Others weren't prepared to accept the second part of that assertion, the notion that moral hypocrisy is necessarily enlarged by one's achievements. The student amplified his use of the word "success," bringing to it an interesting subjective interpretation not always discussed in our secular, materialist culture: "It's important to remember, I agree, that Mr. Head is no college professor or CEO; that Hulga isn't a professor, either, or even a cosmopolitan who has really done well in a Southern city, like Atlanta or New Orleans. Mr. Head is an old man from rural Tennessee, I think; he's been around, we learn, and knows something of city life, but he's *not* one of O'Connor's sophisticates; and she knew how to give us one of them, when she wanted—like Julian in 'Everything That Rises Must Converge,' or Asbury in 'The Enduring Chill.' Hulga is a snob, but she's a little too coarse to be up there with Julian and Asbury, and she is, remember, deformed, and has taken a deformed name, Hulga. [Her real name is Joy.] She and Head are both 'country people,' and yet they're teachers—the best teachers O'Connor could find to lead the class on the subject of pride and vanity and 'phoniness.'

"I guess she was trying to remind us that to hit at the easy targets—I mean, easy for yourself—is to strike out. That's why she didn't only go after the 'interleckchuls.' She spent her time (as a writer) with the people her Lord did, the common folk, and she found in them all the sins she had been seeing among the gentry, her family and neighbors. I guess I'm saying that there too [among the common folk] you'll see successful men and women. I remember when I was a boy hearing people called 'big shots.' It wasn't used to describe people who were rich and famous, not in our poor little world! It meant someone who tried to give the impression he was really important—that he knew a lot, that he was someone you should look up to, that he mattered. 'He thinks he's a big shot,' and then I'd hear, 'and he's a fool, really'; or I'd hear, 'and he's a horse's ass, if you look real close.' Once I asked my mom how so-and-so got to think he was a big shot in the first place. She said a person might do something good and impress people, and he won't let it all rest there. He'll be hungry for applause, and when he gets it, he isn't filled up. He wants more and more; he's never satisfied. So he becomes Mr. Big Shot."

But what might a reading of O'Connor stories do to take the wind out of a big shot—us, for instance? The question, asked by a student sitting right next to the one who had spoken and right across from the teacher, was not meant to be ironic, though we all, I think, recognized the danger of becoming self-congratulatory critics of those who are inclined to self-congratulation. The student wanted a specific exposition of what literature offers the moral life, meaning our moral understanding of things, and meaning, above all, our moral conduct. None of us was able to produce, on the spot, a confident statement of cause and effect: one reads certain novels, and one becomes this or that kind of person. But a hitherto quiet student ventured into this most important and hard-to-survey territory with a succinct introductory foray: "Just to point out the irony is a start. Just to mention that Flannery O'Connor and us the readers—that we can be as smug and blind as the characters in her stories—is to take a step. I don't think we can stop there, though. 'Look at us, what we are doing in this class, the way we're talking and confronting ourselves!' We could talk like that and give some other novelist a great chance to zero in on us, to mock us, and we'd deserve it: the smugness of those who have a good nose for smugness!"

He stopped there, leaving all of us also stopped—in our tracks. After a long silence, we gradually mustered a willingness to take chances. We pointed out to one another that a story is not an idea, though there most certainly are ideas in stories; that reading a story is not like memorizing facts. We talked of the mind's capacity to analyze. This capacity—to abstract, to absorb elements of knowledge, and to relinquish them in statements, verbal or written—is an important part of what we are: creatures of language, of symbols galore. But we need not use ourselves, so to speak, in *only* that way. We have memories; we have feelings. We reach out to others. We have the responsiveness that one sees in preliterate infants who cry when others cry, smile when others smile, frown when others frown; or the responsiveness of youngsters, even preschool ones, who sing in response to the sound of others singing, who get choked up when shown a sad picture or told about a sad event. That side of ourselves is not set apart from our intellect. In order to respond, one remembers, one notices, then one makes connections—engaging the thinking mind as well as what is called one's emotional side.

How to encompass in our minds the complexity of some lived moments in a life? How to embody in language the mix of heightened awareness and felt experience which reading a story can end up offering to the reader? Not that a novel cannot, also, be an occasion for abstraction, for polemical argument. But it can, as well, insinuate itself into a remembering, daydreaming, wondering life; can prompt laughter or tears; can inspire moments of amused reflection with respect to one's nature, accomplishments, flaws.

· · ·

"Who is this Stecher?" a student of mine at Harvard Business School asked, referring to the central figure in William Carlos Williams' trilogy, an immigrant who eventually rises high socially and economically. The student—to our fascination in that class—was intent not only on an intellectual summary. "He's a mass of words, I suppose—what we read in a novelist's book. But to me Stecher is—oh, now, part of me! What do I mean? I mean that he's *someone;* he's a guy I think of. I picture him and can hear him talking. And he looks different to each one of us, and the way he talks is different for each one of us, because each of us has our own accent. He's inside us, and so is Gurlie [his wife]. I can see him walking, or working, or climbing those stairs to his apartment, or eating—and listening to his wife, as she pushes him to get out there and climb up the ladder. Williams' words have become my images and sounds, part of me. You don't do that with theories. You don't do that with a system of ideas. You do it with a story, because in a story—oh, like it says in the Bible, *the word becomes flesh.*"

The student had more than made her point; she had described the consequences for each of us of reading and responding to a story. Its indirections become ours. Its energy invites our own energetic leap into sadness, delight, resentment, frustration. Psychiatrists use words such as empathy, identification, introspection. But in the end the explanatory formulations are not as helpful as what the student said after she had quoted the opening of the Gospel according to John: "I can close my eyes and put cotton in my ears, but I still can see the Stechers. In *The Trial* Kafka can reduce his character to the name 'K,' make him anonymous and mysterious and elusive—but not for us readers: K becomes what we bring to him, and what the story brought to him. We see him. We imagine what he looks like. We put ourselves in his shoes. He lives in us—and he might even make a difference in my life, because he's part of what it is that's in my head that remembers and decides and is loyal to this and doesn't like something else. He is part of my mind's life."

Another student wanted her to define the phrase she used to allude to her mind's life. But general laughter greeted the request. We had about worried the subject to its and our exhaustion. We were left with our sense that Dr. Lydgate and lawyer Jaggers and businessman Stecher and Ibsen's "master builder" don't only occupy lives inside books, but live in countless minds. The whole point of stories is not "solutions" or "resolutions" but a broadening and even a heightening of our struggles—with new protagonists and antagonists introduced, with new sources of concern or apprehension or hope, as one's mental life accommodates itself to a series of arrivals: guests who have a way of staying, but not necessarily staying put.

·6·

The Private Life

STUDENTS OF ALMOST ANY AGE make the distinction between going to class and living their lives, as if all that time in "home rooms" or lecture halls isn't a real part of their existence. But I have heard many others, especially so-called working-class people, make a similar distinction—between their lives as they are spent at home, with family and neighbors, and the "time" (the word can be ambiguously used, sometimes connoting a term in jail) they spend at work.

Even members of the *haute bourgeoisie* make that second distinction. An already successful tax lawyer named Harold, thirty, spoke to me about his work and his "several loves," as he kept calling his family and certain intellectual interests of his: "I was always smart. In the second grade they wanted to have me skip a grade. My mother said no, but she warned me not to sit on my laurels. What are laurels, I wanted to know. She told me. She was always telling me to do better and better. My father was very smart, but he never amounted to much. I think my mother was afraid I'd be a chip off the old block. He was a patent lawyer, but he was also a socialist. Some combination! He would make a hunk of money on a case; and then he'd get depressed. He'd sit and stare out the window. He wouldn't eat much. He'd sneak drink.

"We never had the stuff in the house, and only when I was

sixteen or so did I find out why. I was looking for my lacrosse stick in the cellar of our house. I'd forgotten where it was. I saw a trunk, and on top of it were some of our old skates. I started looking at them, remembering how I learned to skate on a pond near our house. The skates seemed so small! I decided to open the trunk. I wondered what was in it. I can still remember the moment—seeing a bottle of whiskey and a bottle of vodka in a brown bag. A strange place for them to be! Something clicked—memories of my father going down to the basement to lift weights, and coming up 'dizzy.' My mother was always telling him that weightlifting was no good for him; but he said it was the best exercise in the world, and it helped him relax. He'd come upstairs and go right to his room, to change clothes.

"I felt I had suddenly become a detective. I went upstairs and started going through my dad's closet and his chest. In the chest, between the shirts, I found a small, pint-size bottle. It was almost empty. In his closet, in the pocket of one of his sports jackets, I found another bottle. It was whiskey—bourbon or Scotch. I stopped with that second discovery. I felt ashamed of myself all of a sudden. I was spying on my father! Why? Who was I to do that! We were reading *Catcher in the Rye* in school around that time, and I remember thinking: my dad—he's one person who *isn't* a phony. That was his problem—that he couldn't pretend like other lawyers, that he saw what was going on, the greed and the schemes. Boy, was that his word: 'Schemes and more schemes,' he'd say. My mother never stood by him when he talked like that. She shrugged her shoulders. 'That's life,' she'd say. She wanted him to make more and more and more money. They'd fight, terrible fights. 'You want me to be a schemer,' he'd shout at her. 'You bet I do!' she'd shout back. She ran up these bills. What the hell could he do but go and get a case, and win it, and make a hunk of money, to pay off the bills. Then he'd get low, so low you'd think he was going to die. He'd stop eating. He'd stay in his room, and I'd hear him crying. He'd go to the movies a lot. I remember him going to film festivals, instead of to work. He liked Ray Milland and Bette Davis; he liked the

Hitchcock movies; he liked old movies from the thirties that they brought back to this 'art theater,' the only one in the city.

"He was, I now realize, an episodic drinker. He'd 'binge out' — the expression he used when he finally talked about his drinking. But suddenly he'd snap out of it and go dry for months. No A.A. No shrink. No support group. No medicines. And a week or so before he started drinking he'd prepare himself for it, and he'd let my mother know. He'd tell her he'd paid her 'god-damn bills,' and fixed everything in the house that needed fixing, and straightened out his will, or his taxes, or whatever needed straightening out, and so he was ready to 'relax.'

"That was the code word. We all got to know it. We dreaded it at first; but I have to admit, as I got older, and learned more and more about my mother and my father—about their values, about what each of them stood for—I began to understand why my father boozed it up, and I was secretly glad when he started drinking, because he said things then I've never forgotten. He let loose—and not like a drunken fool. He talked about the 'crooks' in the firm, and the 'deals' being 'pulled' all the time, and always the 'schemers' and their 'schemes.' He was an old-fashioned populist, and he hated stuffy, big-deal, uppity people. My mother was always trying to work her way into higher and higher social circles. She was a name-dropper. She was a very bright person. It drove her crazy that my sister wanted to be a nurse. A nurse! My sister is my dad's girl. She wants to help people. She joined the Peace Corps; she worked as a volunteer in a hospice. My mother has never given her a moment's peace. She likes me because I went to Harvard, and Harvard Law School. She doesn't listen to me talking about what I believe. So long as I make money in the six figures, and live where I do, she can talk about me nonstop with pride. When she comes out to see us, I keep visualizing those bottles I found hidden when I was a kid, a teenager; but I'll never drink the way Dad did—and I'll tell you, I'll never be as honest he was!

"Don't get me wrong: I'm no crook. I'm just your average up-and-coming corporate lawyer, with an additional boost because I took the combined Law School and Business School course, open only to a few of us real ambitious characters! Psychiatrists would say I've got plenty of my mother in me, though some of my dad. My wife is my shrink. She majored in psychology, and she's got a lot of gut intuition besides. When my dad went on his last binge, she was afraid it would kill him. She told me, 'He wants to die.' And he did die. He got in his car, and he drove that car into the wall of a garage. They thought he must have been going sixty miles an hour. What a way to die.

"My mother—all she was worried about, when she got the news, was how the death would be written up. When it was called an accident she relaxed. She never really cried. She was relieved. He'd left her a couple of hundred thousand dollars and a mortgage-free house. Her brother is loaded. He helps her. She's happier since Dad died.

"I'm already on an escalator, a fast-moving one. I'm going up! But I have them [his colleagues at the law firm] worried; every once in a while, I tell them I've got to get away. I think at first they thought I was an alcoholic. But I'm a gambler, that's what. No, I don't gamble. [I had asked.] I mean, I gamble in a different way. I know my work. They know I'm good. I'll feel things tightening all around me. I watch people and they seem like plastic puppets, pulled in all directions by the market, by the firm, by their own greed: they want it all, and the sooner the better. For a stretch I'm in there with them. I plug along and come up with these bright ideas, and my sponsor [the senior member of his firm who is his mentor and friend] says, 'Great, great!' But then I'll wake up one morning and feel like a dead man. I feel as hollow as one of T. S. Eliot's 'hollow men.' I'm ready to fold; I'm ready to go running and never in a million years come back. And that's when I throw the dice; I mean, I tell them at the office that I'm not feeling well. I call in.

"I stay home. I just read and think and take my walks. I go to the library and sit and read. I look at the people in the library. Who are they? Ne'er-do-wells? Housewives and their kids? The elderly? Those ways we have of talking about people! I used to hate to go to the library at Harvard. I'd see everyone grinding away, and I'd feel so damn competitive. I'd hate them; I'd hate myself for hating them; I'd hate my mother for bringing me up to be the person I am. I remember thinking: go and get *War and Peace*, or any really long novel, and just sit there in Widener [Library] and read it, and let all your lousy friends, with spikes in their elbows, see you, and let them smile and think their 'chances' are better, and *let* their 'chances' be better—to hell with the whole damn scene! But I couldn't; I'd 'psych out.' I'd go back to cramming, and only when I was back in my room and in bed and waiting to fall asleep—only then did I think of my father. I'd remember his speeches, the way he talked about 'conniving schemers.' I'd remember him telling me to read Shakespeare, and to hell with all those economics courses I was taking. 'What's happening to you?' That's what he'd ask. He'd have had me become a Shakespeare scholar. He was one. With drink in him, he ranted and raved his Shakespeare soliloquies that he'd memorized as a boy and went back to while drunk.

"Well, you see the split in my life! But I don't drink; I just go on book binges. I pull out of the whole damn rat race, and I sit and read at home and at the library, wherever I feel better. Sometimes it's nice to have people around, so I'll go to the library. I'm going through reading lists I picked up in college: courses I didn't take, courses I should have taken, and wanted to take—oh boy, did I!—but I didn't take them. They wouldn't help me with my career. They weren't as 'important' as others were! My mother would brag about me with her friends, but never to my face. She was a driver, a slave driver. She's old and sick now, but I know her as she was when I was a boy. She never read Shakespeare and thought novels a waste of time. She followed the stock market and read *Time* and *Newsweek* from cover to cover—and like my wife says, she was a great one for 'the reality principle.' She was always saying, 'Let's be practical.'

"You see what I mean—the gambling I do. It's tremendous fun! I'm ready to be fired. Let them send the word down. I conduct my mini-challenge—to myself. I guess I'm being childish. But there's part of me that's not ready yet to surrender. Others get the flu. I get my urge to go sit and read. I'm not 'sick,' so far. I'm discreet—'cautious, politic, meticulous.' I remember those words in one of Eliot's poems. I try to be honest. I tell my friend, my mentor, that I'm strung out and I need a couple of days to get my bearings again. He says fine. He says it's better I go and read than get drunk or use drugs or fake sickness. And I could never leave when there's a real jam at work. I'm a safe rebel! Maybe I'm just kidding myself and no one else; maybe I'm no rebel at all—just a very efficient, hard-working guy who knows when he might blow and has figured out a way to keep himself on track.

"I'll say one thing, though: without those books I'd be locked up someplace, either a jail or a mental hospital. I'm too sensitive, I'm afraid, not to be troubled by some of the things I see around me on Wall Street and in the law offices that keep Wall Street going—but I try not to think about all the sleaze, and I do the best job I can. So far, I've not had any personal contact with the sleazy types. So far, I can call myself clean. But for how long? And what would I do if I were caught in a tempting situation? Would I rationalize my way into becoming a card-carrying member of one of the local sleaze groups here? It's questions like those that get the better of me. It's questions like those that get me into a panic—and that's when I end up reading in the library for a day or two."

For him the library was a sanctuary, a place of moral refuge. Harold was *not* a psychiatric patient, though he had the wisdom and wit to realize that his "escapes" (as he occasionally called them) could easily be called neurotic by some doctor. "I know I have troubles," he once said, in case I might be tempted by clinical labels, "but I'm surrounded by troubles, and a lot of them are questions of right and wrong." I had known him as an undergraduate, then as a law and business graduate student. I knew that he took literature seriously; he majored in it at college, despite the abundance of economics courses he took, and he continued reading novels and short stories and poetry during his graduate school years. In my "Dickens and the Law" course, he sometimes discussed the character Wemmick in *Great Expectations* — the lawyer who was thoughtful and generous as a husband and a father but quite another kind of person when at work in the lawyer Jaggers' office. This split behavior, this schizophrenia of sorts (not the kind we study in psychiatry and psychoanalysis), struck him as odd—until, of course, he matriculated into the ultimate "learning environment." In a sense, Harold would never let Wemmick depart his mind. He once joked with me, during a phone conversation, that Wemmick was his "patron saint."

Like others in that class and in other classes I have taught, Harold wondered how to straddle different worlds, and, too, how to make sense of his own behavior—the different responses his mind made to a private life and to the so-called professional life that takes place in offices. Harold was eager, also, to contrast Wemmick with another of Dickens' lawyers, Sydney Carton. *His* private life was a shambles—heavy drinking, loneliness, no family, a skeptical if not cynical nature. Nor was he a successful lawyer. He helped the aptly named lawyer Striver, but seemed as defeated professionally as personally. Still, Sydney Carton is the hero of *A Tale of Two Cities.* Harold was one of Carton's admirers, though perhaps it is better to call him a critic of Dickens, or maybe a doubter with respect to Carton's plausibility. That skepticism became, for a while, an intellectual obsession for him. He went to see the movie made of the novel, in which Ronald Colman stars as Carton. He reread many of the novel's passages in an effort to comprehend both the character and the author's intention. He even took the book with him to the office, hoping to pick it up in spare moments.

It was this step that prompted anxiety in him, enough of it to make him wonder whether "trouble" wasn't "around the corner." When I asked him what kind of trouble he had in mind, he smiled and insisted that he was only being "ironic." He spelled out the dimensions of the irony in this way: "I'm no saint, and I'm not trying to be a saint. But I *am* worried about becoming something worse than a sinner—a guy who has lost his integrity. I feel myself slipping into bad habits—the way I treat people, and the way I'm starting to rationalize what we do in the office. I've not violated any laws, only a code of behavior which used to be a law for me: that I be polite and considerate of others, and that I not be used by people whose values I don't respect. I'm being used by anyone who can purchase my time—and to be honest, some of my clients are awful people. I believe they're guilty, and I'm trying to secure 'innocent' verdicts for them, and I don't like some of my tricks. I've become a real sharp lawyer!

"What would Carton do? That's my 'grand obsession,' my big question as I tiptoe about, high up in that skyscraper, looking at Manhattan and the river and the cars fighting it out. Carton would quit. Carton would throw a fit. Carton would go and get so drunk he'd never wake up. My colleagues, a lot of them, they get plenty drunk: yes, plenty drunk, but discreetly drunk, most of them. They're not Carton, moping around and feeling scornful about a corrupt world. They are strivers, like Carton's employer, Striver, and they drink, *we* drink, because we're eagerly in a rat race, and every once in a while the tension gets too high, and you need to 'blow off steam.' I hear that phrase all the time! I wonder whether Dickens ever thought of *that* side of law: the need it inspires to 'blow off steam'!

"Maybe you're right. Maybe when people are troubled about what's right and wrong they do get tense, and so they drink, or snort coke. But I sit and hear the folks talking at the bars, and they're not feeling guilty; they're in the fast lane, and they're fueling up, just to stay there! I hear them say as much and I hear *myself* say as much! I can hear Carton, though—*my* version of Carton, not Dickens' version. Do you know people who do that—take those characters in the novels you teach and turn them into buddies of theirs? Soul mates? I'm not kidding! And I'm not—I hope—hallucinating. I can hear Carton talking, telling me what he thinks of Wall Street law. The Carton in my head is a mixture of the Carton in *A Tale of Two Cities*, and an uncle of mine—the ne'er-do-well in the family, who taught in a junior high school!—and my father, with his frustrated idealism, and my own self. Sometimes he'll be Dickens' Carton, giving me a formal speech and telling me that I'm being a greedy opportunist and will pay for it some way, someplace, somehow. But then I'll hear my uncle or my father telling me you only live once, and you should do what you believe matters, what you believe in, and not tie yourself to money and a big-deal life. It's not my uncle, though, and not my dad. It's Carton talking like my uncle or dad did. Then there'll be me, plain old me, talking to myself, the way I always do—but I swear, I feel like I'm Carton!"

We discussed this psychological phenomenon at great length. I told Harold that other students have described a similar psychological event: a story's character becoming imbedded in their mental life. Characters who have taken a special hold include some of those who people Walker Percy's novels, such as Binx Bolling, the protagonist of *The Moviegoer*, and Will Barrett, a central figure in *The Last Gentleman.* Binx is a twenty-nine-year-old stockbroker who is trying to figure out what this life means, how to live it, the old existentialist inquiries. His shrewd observations, his satirical broadsides directed at all sorts of American habits and values, his lively sense of humor, only partially mask a kind of moral despair that won't let go of him—a revulsion, really, at the hypocrisy and greed Sydney Carton and Harold couldn't stop noticing. Will is a somewhat vague but intensely curious person who lacks Binx's acerbic wit, perhaps because he is far more innocent and vulnerable. As he moves across America, north to south and then west, he engages with others who in various ways remind us of ourselves, contemporary men and women going through our motions: doctors, business folk, people interested in religion, sick people, dying ones, the young and the aging. Will is trying in his earnest, open way to fathom all of them, to understand what it is that keeps them going about their appointed tasks. His almost infinite patience with such fellow creatures is remarkable, and has Christian overtones. Both Binx and Will are young—the age of many graduate students—and their search for meaning, their rueful look at twentieth-century

America, engages university students. But both men, Southerners, are not at school; they are living their lives, one, Binx, as a stockbroker, the other, Will, as (for a while at least) a "maintenance engineer," meaning a janitor. Will had, actually, left Princeton without graduating, worked briefly as a law clerk (many of his ancestors were lawyers), served in the army for two years, learned more than a little about electronics, and ended up in Manhattan taking care of a building while seeking psychoanalytic treatment. The two have in common their post-academic youth, their wide-eyed openness to the world's possibilities, counterbalanced constantly, however, by a wry and sometimes brooding sensibility that prompts them to fend people off, be doubtful of much in the way of habits and values all too readily embraced by others. Will is explicitly described as a wanderer, a watcher, a listener; Binx Bolling lives up to his name, at least for part of the time—someone rolling along, as hard to catch and hold as a slippery ball. Neither of them is ready to settle into life's conventional grooves, though both are hungry, underneath their façades, for a life that connects to others, one that makes moral sense.

Men and women who return to school in their middle or late twenties or early thirties seem especially responsive to *The Moviegoer* and *The Last Gentleman.* Business school students, young doctors, and working people in their thirties who have taken a course I teach at Harvard's Extension School have all found much in these novels that speaks to them. They may have decided what occupation to pursue but have not made their way into a personal life that they find worthy of respect and commitment. As some class members openly speculate, why, unless there were something missing from their personal lives, would they even consider taking a course devoted to novels instead of one of the more "practical" courses that beckon?

The theme that emerges is mild restlessness or apprehensiveness occasionally turning into gnawing discontent, agitated despair—and drinking, extramarital romancing, trouble "settling down." Percy's novels and John Cheever's short stories are especially helpful for readers struggling with that sense of restlessness. Gin and bourbon, a fling or two, one purchase after another, new and more esoteric vacations, a second or third home, the suburban scene, the country scene, the seaside scene, bashes and teas and lunches and wedding receptions and celebrations of anniversaries, birthdays, graduations, promotions—and still a certain emptiness and confusion and even sadness persist.

No wonder Janice, a twenty-eight-year-old survivor of all that, ready (by the second year of Harvard Business School) to start in all over again, expresses the gravest reservations even as she prepares to return to an important New York City brokerage house: "I read Walker Percy's novels, and I get impatient with him. His women are all vague or badly hurt. His men interest him more, but they're also vague or hurt, at a loss to know what to do. It's scary. I want to put the novels [*The Moviegoer, The Last Gentleman*] aside, but I can't. The reason? Because the people in them remind me of people I know. My cousin says Percy's characters are Southerners and you have to be a Southerner to really understand what he's getting at, but she's not a Southerner, and the way she talks about those characters strikes me as pretty good—I mean, she seems to be counting herself in as an honorary Southerner. I don't think the South is what Percy's describing; he's describing my cousin, and her brother, and me, and my two brothers. He's describing lots of us on Long Island and in Connecticut and the suburbs of Philadelphia and Boston.

"We put on a show—we think a show for others, but really a show for ourselves. We kid ourselves. We get rid of ourselves. We disappear into movies and musicals; into condominiums and mansions; into specialty shops—lost in our lust for clothes, food, newfangled appliances. The other week I went home, and my mother had just come back from the shopping mall. She's always going there, and always coming back with something. We only see *most* of what she brings back. She hides the hootch, and she hides some of the jewelry or shoes, and only gradually introduces us to what she owns, what she's bought on her various trips. Everyone laughs at Imelda Marcos, *her* shoes; but I'll tell you, when I sneak into my parents' bedroom, and look in my mother's closet, and see all those shoes of hers neatly stacked, I want to cry.

"I remember when I was eleven years old. I was in the sixth grade, and she took me shopping with her. She didn't know what to buy—that's what she told me as we drove there. Then I asked her, innocently, why we were going. The next thing, I heard her sobbing as she drove. I saw the tears on her face. I didn't know what to say or do. I decided not to say or do anything, I guess because I was afraid we'd get into an accident if she got more upset. She pulled a handkerchief out of her pocketbook and wiped her eyes. I sat there, silent. Then she started speaking. She told me that 'you can't have everything in life.' I can hear her saying those words, right now. Once again I didn't know what to say. Ask her what she meant, because I wanted to know? Agree with her, even if I didn't know what she was getting at? I just sat there—and she went on, pretty soon. She told me that Daddy goes away a lot (as if I didn't know!) and that she used to miss him, but she'd get very 'low,' missing him, and want to sleep all day. So, one day, she snapped out of it. Her sister told her to go and make herself happy, by buying something for herself, and she decided to try the advice, and she did, and she felt much better, and that was that. She told me she'd been buying things ever since! I remember, many times, thinking my mother was really having a good time for herself, and I was glad—but also thinking that there was something wrong with this person who was sitting or walking beside me, or driving that Oldsmobile station wagon, I think it was. (She's got a Mercedes station wagon now!)

"Back then, when she told me about why she went on buying sprees, it was the first moment in my life that I felt distant from Mom, as if I was someone from another part of the world, or from a completely different world, who was watching her and trying to figure her out. When we parked, I remember seeing her eyes: they moved from store sign to store sign. I felt, inside me—I had no words for it then—her hunger. She wanted to consume something. She told me we should 'take a walk,' and we did. She would hold my hand now and then, but I could have been on Mars for all she cared. She was trying to figure something out—where to land, where to strike. Yes, there was a bit of the predator in her! She was circling. As I said: she was hungry—a bird of prey!

"That was when I first saw her buy a pair of shoes. She wanted the most expensive kind. The man would bring out one pair, and she'd look at them and ask how much, and then ask to see other pairs, and the same thing happened, and I was just sitting there and looking at her, and at that man and his shoehorn and his boxes piling up, with the shoes half put back in them, and my mother, staring at her feet and rubbing one foot against the other and pulling on her hosiery, and every once in a while, she'd ask me if I wanted anything, meaning: go next door, where there was an ice cream parlor, and get yourself a cone or a soda or a sundae—'Anything you want,' she added. Those last words were meant for her own ears, I realize now. But I didn't leave; I kept near, and watched. After a few times on those sprees, I had the whole routine down cold in my head. I remember an antique store, a ski shop, or a department store, where she'd buy all those perfumes and colognes, whatever it was she wanted when we got there.

"The worst times we had were when she couldn't make up her mind where to go—maybe because as much as she tried, she couldn't think up something to want, to go buy. Need was beside the point; she had everything. Desire was the point—to think of something she wanted, and to 'go for it.' I remember one Saturday, we went to the mall, and she looked and looked, and I could see she was getting lower and lower. She started getting irritable with me. I knew what she was thinking: she wanted to take me to the ice cream place, then tell me to go sit in the car, after I'd had my lunch. I said I wasn't hungry. Then she got her idea: we should both 'go have a really nice lunch.' Oh oh, I thought to myself: what would happen now? She went and looked in the phone book. She took me to the nice restaurant at the mall, but then she decided it wasn't 'nice enough.' So we went back to the car, and she drove us to a real fancy place in a town about twenty miles away—all for some food! I got fidgety; I was nervous. I was a kid. I wanted a cheeseburger, some fries, a milkshake—and she's driving and driving. I could tell by the *way* she drove that she was in a bad mood. She was speeding, then she slowed down so much, people would honk. She was in the outside lane, but she wouldn't move over. My poor mother! We got to that restaurant, and she perked up. She asked for a place near a window, one of the best seats, so we had to wait. I really did want to eat by then. I told my mother so, and she got very upset with me. She sent me to the rest room to 'freshen up.' I stared in the mirror and came out. I felt real lousy!

"Why am I describing that day, the lunch? It was a terrible time. My mother ordered all these fancy dishes, and then she didn't want to eat any of them! I knew her tricks—moving food around on the plate, and eating a little, but leaving most, while saying she shouldn't have had such a big breakfast (or lunch), even though she'd probably had black coffee for breakfast (or a salad for lunch!), or saying she felt 'a little sick.' Did I want to nibble on her food? No way! I knew, when she ordered, that she'd never eat those dishes. They were fattening. They were what she really did want to eat, but never let herself eat. (Later, a year or so, I discovered her secret: she not only hid bottles of vodka, she hid bars of chocolate!) When we left that restaurant I could see that Mummy was really unhappy. I wondered what she'd do. You know what—she drove back to the mall and (you guessed it) bought some shoes. She perked right up! I didn't know whether to laugh or to cry. I was about thirteen.

"No wonder my mother watched those TV soaps. No wonder she read all those stories in *Redbook* and *The New Yorker.* No wonder she'd go to the movies—I remember her seeing *I'll Cry Tomorrow* three times—and come home with her eyes red, and she'd go upstairs to 'freshen up,' and come down afterwards with her speech slurred and walking wobbly! By the time I was fifteen, maybe sixteen, I knew it all: her drinking and her loneliness and her spending sprees, and Dad's 'friend' in San Diego. He was always flying there on business, and always having to stay an extra day, or the flight back got canceled, or he missed it. I think my mother caught him—somehow she did—and by then she wasn't even surprised or upset! I never heard them argue. I think she probably got drunk and then went on a buying binge. Once I heard her telling my father 'never to embarrass' her, and I figured that they were talking in their code language: she hinted, and so did he. Maybe I don't know everything, though. Maybe they figured out a way of fighting without me knowing. All I know is that when I first read those Cheever stories, I thought he'd been spying on us. When I read Percy's novel [*The Moviegoer*] I wished my parents could read it. It's too deep for them, maybe, even if they did go to good colleges, though I think they'd be able to get the message, if they ever wanted to work at it. But they're each in a rut, the biggest rut of them all, the kind of sad life they live, supposedly a life together, and supposedly a rich, successful life. It all makes me want to go study turtles in the Galápagos Islands, the way my college roommate did!"

Stories such as Cheever's "The Housebreaker of Shady Hill" and "The Sorrows of Gin" have a way of working themselves deep within the thinking of many students. Some of them, even a year or two afterward, in letters or visits, make mention of what it still means for them to think of the girl Amy and her family and its maids in "The Sorrows of Gin," or of Johnny Hake and his wife and their neighbors in "The Housebreaker of Shady Hill." The latter story, in particular, is explicitly moral in its central preoccupation. A thirty-six-year-old prospering businessman has had a financial setback. He is far from starving, but he is short of cash and beginning to wonder about his prospects. In the middle of the night, he finds himself drawn to a neighbor's, a friend's, home—to his wallet, which (with its nine hundred dollars) he takes home. Now the worried man becomes a tormented one. His conscience asserts itself gradually, affecting his capacity to trust and believe in people, his own wife and children included. He becomes isolated, brooding, a victim of his own deed, his own self-knowledge. He also becomes, for many readers, an interesting companion of sorts, a stimulus to all sorts of reflection—with respect to not only his deeds in the story, but those of others who live in Shady Hills across America.

Cheever addresses the questions of crime and dishonesty with a brilliantly suggestive concreteness. Johnny Hake is no demented crook from whom we may easily distance ourselves. He is charming, well educated, and of "good" or "proper" background; he speaks to us directly and wins us over through the ironic humor with which he describes the privileged world he is part of. Cheever's fondness for Johnny Hake, for his wife, and for their way of living is evident, even as he gently though insistently portrays its hypocrisies and banalities. The story offers its fair share of elegantly written satire, but it works, additionally, a magic spell of wistful yearning on many of us as we recognize in Johnny Hake's dilemma a symbolic representation of our own continuing struggle with ambitiousness and greed. His deed is notably a private matter—never an official crime with police involvement. To some extent the issue is his culture's reticence and gentility—a hesitation to call the police when a wallet disappears from a house not broken into. But Cheever clearly wants us to approach Johnny Hake less as a sociological specimen than as a moral protagonist.

Students have told me for years that they can't get Johnny Hake out of their minds. Weeks after reading the story and dutifully discussing it in class or writing an obligatory paper on it, they are still "picturing him," as a student named Michael put it during an office-hours discussion. I was interested in the way he visualized things, and said so. He explained: "From the very start of that story, I started picturing him, *seeing* him—because that's what Cheever wants, I think. He gives you his height and weight, as I recall, in the very first sentence—no, the second. 'My name is Johnny Hake' is the first, and then comes his age [thirty-six] and his height and weight. I'm five eleven myself [Hake's height] and so I latched on to that, and I'm in my 150s, a little bit heavier than thin Johnny [whose given weight was 142 pounds]. But he was 'stripped,' I remember, when weighed; I remember that because I remember my own father taking off all his clothes before weighing himself.

"Johnny played football (and I think baseball) in Central Park. I've never done that, but one of my [prep school] roommates did that all the time. I tried to get him to read that story, but he wouldn't. He said, 'You're a bad advertisement for Cheever.' Why? Because he thought I was 'on a downer' after reading the story. But he was wrong. The story didn't depress me. It made me stop and think. To answer your question—it made me see my life, my dad's life, like I'd never before seen anything. Especially that scene of him going into the Warburtons' house; you don't just read those lines and think of Johnny becoming a house robber. You see him walking into a house he knows from many a cocktail party, and going right up the stairs, as if he's got to go to the bathroom and the downstairs bathroom is taken and he just can't wait. Sure, he's nervous, but not *that* nervous. He gets there. He lifts the wallet. When the wife wakes up for a second and talks to her husband—well, no panic for our friend Johnny Hake!

"It was later that he began to crumble—I mean, become tormented. Then, too, I *see* him: raking in the dough on Sunday in church [he helps take the collection], and reading the paper and noticing how much robbers are everyday news. I remember staring at those ushers in church, wondering who in hell *they* were to be taking in all that money, and who they were to be making me feel cheap or stingy because I didn't give as much as I thought I should, or as much as I thought they thought I should. Boy, do I remember the tricks we pulled, my father and my brother and me: we'd push our hands with the money into the plate so no one could tell how much we were giving, because our money would end up below the rest of the money. We'd hold the money in a fist and then release it, spreading our hands over the entire plate as we did. I remember those times—secretly wishing I could *take* from the plate, not put bills into it. Sometimes we'd put in coins, not bills. Then our job was to do it without making any noise—the coins not tinkling or rattling, or else people would know that we were cheapskates. My father would get down on himself and call himself that—a cheapskate. Other times, he'd be like Johnny Hake—tell us about all the crooks in the world.

"Johnny suddenly noticed how many crooks the papers wrote

up after he'd become a 'housebreaker.' I don't think my father ever did anything like that. What he *did* do—well, he owned a lot of property, and he pulled in a lot of rents, and many of his tenants weren't all that well off. I remember him on Sundays, early each month, figuring out which tenants still owed him dough. I was amazed at all the money he got from those people, but I think he had a lot of trouble with himself—the idea of profiting from others who weren't in the greatest shape. I was a kid; I was no psychologist, but I remember noticing that Dad said some pretty bad things about his tenants around the time he collected their rents. I thought of him and his business when I read that story."

Students who have little in common with Johnny Hake's world nevertheless find Cheever's story powerful—especially those who aspire to live someday in one or another Shady Hill. A white, working-class student of mine named Jack had been introduced to the story in high school by an older brother, Ken, an accountant who was earning a law degree at night. Jack found the story "sad," as have many others who have read it, but he was interested in Ken's response: "He's married; he has a kid, and another on the way. He dreams of a rich life, though he tells me he'll settle for one 'fifty percent better' than the one we've got. Why not one hundred percent? I've asked him that a lot, and he says he doesn't want to lose himself. I couldn't figure out at first what he meant. Then he gave me the Cheever book and said I should read 'the Shady Hill story.' So I read it, and it didn't mean much to me. I thought Johnny Hake was a jerk, spoiled and full of himself. I was a high school kid—a bit spoiled myself, and as my brother once put it, 'with a high opinion' of myself. He wouldn't let me forget Johnny Hake. I think he got scared, reading that story. We have long talks in class about *Crime and Punishment,* but for my brother that Cheever story was enough to get him thinking the way we do here in school. I think he needed a reminder of what's really important in life, and that story gave it to him. I couldn't figure out why for a long time. It'll take a couple of generations for us to get that far up the ladder. But when I read that story a second time I saw what Cheever was trying to say. Johnny Hake is twice a housebreaker. He slips into his neighbor's house and takes that wallet with nine hundred bucks in it. (Who carries that amount of cash, even today? It must have been a fortune when Cheever was writing that story, ten or twenty years ago!) But the result of one 'housebreak' is another. Hake's conscience won't let go of him, and he makes his wife miserable. She's ready for a separation. My brother kept telling me that the story is about 'life,' not Shady Hill. I wanted to think of 'class,' of the rich, of all those fancy suburbs people like me dream of living in. But Ken kept telling me it could happen anywhere—and he reminded me of *our* neighbor."

I asked Jack about that neighbor and heard the story of a wholesale greeting card salesman, a man who loved to change his shirt twice a day, took great pride in the new leased car his company gave him every year and in his attractive wife and their two young daughters—and who ended up with a ruined reputation, a broken life, for reasons not unlike those that prompted Johnny Hake to act as he did: "In a story the author can make everything come out OK. In life, things don't get repaired the way Cheever repairs them. It's all very well for him to work that magic—the stroke of good luck that makes Johnny Hake decide he's going to return that money. But most of us don't get a second chance. We're lucky if we have *any* luck at all—let alone luck that saves us from being a lifelong crook. Hake's self-respect is saved, and so he can cheerfully walk the street at the end and be the proper nice guy to the local cops who are patrolling. But when our neighbor, a really kind guy, a generous man—he was always buying presents for kids in the block on their birthdays—got caught shoplifting, he was ruined. There was a story in the paper because the store owner went berserk and pulled a gun and threatened to shoot to kill, and a man got scared and had a heart attack. What a mess, over a watch, a lousy watch—and not a nine-hundred-dollar one, no Rolex. A minute in a jewelry store and a family is ruined. Our [salesman] friend loses his job. His wife leaves him. All of us are left trying to figure out why in hell he did what he did. It's not a Cheever story, not as pretty! But there's the same lesson in both—that you cross a certain line and it's a different life awaiting you.

"In class most people said they don't believe anyone would ever do that, sneak into his neighbor's house, not in a Shady Hill! They got bogged down with details: the doors would be locked these days! My father told us that when he was a kid his best friend stole the wallet of the landlord who owned the apartment house where he lived. The landlord collected the rents himself, and he was a fast talker, and he liked his beer, and my dad's friend—he knew how to lift a wallet. You know what? The guy felt bad afterwards. He went and confessed. The priest told him to return the money. The guy said no, he wouldn't; he was afraid the landlord would throw his family out, evict them. The priest offered to return the wallet, no name mentioned. The guy agreed. But guess what? The guy kept ten dollars, a big sum back then. My dad said the ten dollars ruled that guy's life for a while. He splurged; he bought all his friends ice cream sodas and gum and God knows what in the variety store. But he was scared; he was sure someone would track him down. Finally he told his mother, and she got very upset and went to the priest! The long and short of it—the mother gave the priest ten dollars to send to the landlord, and my dad's friend got a second job and saved up the money and gave it to his mother. My father said *his* father hated that landlord and said every tenant should try to steal his wallet, because he's stealing *their* money all the time, but it's 'legal,' so he gets away with it, and the priest doesn't give him a lecture, because that's the way the world is run! I remember asking my father what he thinks of what Grandpa said. Dad said you can't take the law into your own hands. He wouldn't say anything more."

In fact, John Cheever addresses a similar theme in his story. When Johnny Hake returns to his home with Carl Warburton's wallet, he quickly turns on himself and is soon in tears. On his way to work he is full of self-recriminations, calls himself a "common thief," an "imposter." He dreams that "the moral bottom" of his world has collapsed. He now belongs in the company of all those who figure in each day's news, the various liars and crooks and embezzlers who never disappoint the need newspapers have for daily stories. Perhaps a cup of coffee will help. He sits in a restaurant at a table with a stranger and notices him pocketing a thirty-five-cent tip left by an earlier customer. But the author won't let the matter rest there. He has made his point—that many of us succumb to dishonesty or avarice in the small moments of our lives, especially when we think we go unwatched. But he wants to push further; he takes us into Johnny's cubicle at work. The scene is a brokerage house, and the bustle, the importuning, is going on full speed. In this dog-eat-dog world, Johnny's guilt seems oddly attractive, a remnant of some other moment in man's long and troubled history. Even "an old friend," named Burt Howe, is part of this impersonal, hectic, cannibalistic scene. He addresses Johnny with these words: "What I wanted to talk with you about is this deal I thought you might be interested in. It's a one-shot, but it won't take you more than three weeks. It's a steal. They're green, and they're dumb, and they're loaded, and it's just like stealing." Johnny appears to assent, to become a willing, passive accomplice. But deep down he is now an all-too-alarmed moral critic—of himself, obviously, but of the world he inhabits, too.

Cheever is relentless in the scrutiny he gives to Johnny and to his business day. A man who had good-naturedly gone along with a conventional, privileged life now begins to hear and see what once had gone unnoticed. A man in the office next door keeps calling people up and telling each person the same account of what happened on the weekend ("We went out to the shack on Sunday. Louise got hit by a poisonous spider. The doctor gave her some kind of injection. She'll be all right"). The dreariness of it all—call after call, the same words used—comes through to a listener who before might not have even heard the words. Might the "spider" be meant as "a code of warning or of assent to some unlawful traffic"? As Johnny contemplates what he is hearing and what he is tempted to make of what he hears, he tells himself that "what frightened" him "was that by becoming a thief" he now "seemed to have surrounded" himself with others in that mold, "with thieves and operators." But he is not quite right here, as Cheever well knows. Rather, a moral sensibility has been awakened.

Naturally, Johnny seeks relief. In a sentence worth a volume of psychiatry, Cheever reminds us, through Johnny, of what we all tend to do—find someone, somewhere, to scapegoat: "My left eye had begun to twitch again, and the inability of one part of my consciousness to stand up under the reproach that was being heaped into it by another part made me cast around desperately for someone else who could be blamed." In no time he is remembering a troubled childhood, including an episode in which he picked his father's pocket. He is, all in all, a victim rather than a criminal—one more well-to-do twentieth-century American with a past in need of psychoanalytic ratification. But such a discovery doesn't bring him any real comfort. He prepares to go seek lunch, yet speaks to himself of his "moral death." The sky outside is symbolically dark. On Third Avenue he is an odd man out, estranged from a community he had once felt to be his. He searches for decency and goodness, scans faces "for some encouraging signs of honesty in such a crooked world"—in vain. Even the panhandlers seem corrupt. A lunch with his friend, of course, confirms yet again the corruption on the other side of the tracks, his side. Johnny retires to the men's room, sick as can be.

The city seems desolate; but on the train home, that familiar ride from work to the private life, he has a visionary moment, rendered in as beautiful a stretch of prose as Cheever ever offered us: "It seemed to me fishermen and lone bathers and grade-crossing watchmen and sandlot ballplayers and lovers unashamed of their sport and the owners of small sailing craft and old men playing pinochle in firehouses were the people who stitched up the big holes in the world that were made by men like me." Who are those people, and what do their lives have to say to Johnny Hake, to so many of us who sense a connection or two to him? They are ordinary humble people who don't, by and large, live in the Shady Hills John Cheever depicted for so many years. Fishermen don't boast yachts, don't compete in fancy races to Bermuda or elsewhere. Lone bathers enjoy the sea with no desire to exhibit themselves to others. Grade-crossing watchmen do their small bit to make us all live safer lives. Sandlot ballplayers are not in the major or minor leagues. As for those "lovers unashamed of their sport" whom Cheever and his spokesman Johnny have listed on their joint honor role, they are not spoilers. Cheever has never had much patience with puritanism. Those who sit out dances, who find nakedness a cause for shame, who, in Eliot's well-known phrase, don't "dare to eat a peach," are the spoilsports for whom Cheever has little use. As for the men or women who thrill in the wind against the sails of their *small* craft, or the elderly folk who play cards in firehouses—again, they lack pretense, they enjoy themselves in small ways, and sometimes they even manage to help the rest of us out.

Johnny Hake is far from through the ordeal as he grasps for meaning and purpose. He faces further moments of self-absorption, self-pity. But the man is on his way to an important transformation. Cheever is interesting as he ends the story, and my students are ever willing and anxious to discuss that ending, as if (I sometimes feel) their own futures are at stake. "It is not, as somebody once wrote," declares Johnny, "the smell of corn bread that calls us back from death; it is the lights and signs of love and friendship." Put differently, not by bread alone do we muster up the strength to live good and honorable lives. Yet Johnny's friend Gil (whom we met at the beginning of the story as a business antagonist) calls to signal an end to their estrangement. The old boss is dying. Would Johnny now like to return to the job he once held? Only after he says yes, and gets "an advance on his salary," does he return Warburton's money. At night he is now cheerful, able to face the dark, whose significance (in himself, in others) he has realized as never before. The reader smiles, is reassured—but also may not be quite so buoyant as Johnny is.

A twenty-nine-year-old man named Fred who came to Harvard Business School from a well-known midwestern aerospace company loved the lyricism of Johnny Hake's testimony but chafed at his last remarks, which Fred called "gratuitous." He was no fool: he knew that short stories often have to end a bit arbitrarily. Rather, it was a measure of his own immersion in the story that he found its ending hard to stomach and to fathom: "How could Cheever do it? He had me questioning so much, the way Johnny did; and suddenly that fellow Gil picks up the phone, makes a call, and *that* is what makes our hero go back to being his old self. He's whistling in the dark again, as he had been doing all his life: ' "Here, Toby! [Fred is reading the story's last line.] Here, Toby! Good dog!" I called, and off I went, whistling merrily in the dark.' Hey—it's a great scene, and I love it. Johnny Guilt now becomes Johnny Cool; he's at home again, and no cop (or any other symbol of the law) is going to bother him, because he's no longer in exile, you could say. But who got him to shape up? Not his wife; he and she were ready to split. Not his kids; they don't seem all that important to him—like the kids in other [Cheever] stories: the damn adults are so locked into themselves, they have no time for the next generation. And not himself; he's not called by his conscience to go give that money back—not by a long shot. He's called by his friend, the same guy who got him fired at the start of the story. I was disappointed. I thought Cheever ruined a wonderful story. I felt as if I'd been tricked, just when I was on my way to seeing the promised land! Sure, I'm kidding—but I'm not kidding. I was convinced that Johnny was going to return that money, but not by getting an advance on a new and better job. I thought *he'd* become a seaman or a crossing guard. I thought he'd show himself to be a different guy, and that out of that [development] would come his own self-respect; and his wife's respect for him would follow. I thought he'd catch a glimpse of those kids of his, a loving look. Instead—I couldn't believe it: a lousy call from a lousy two-timing *ex*-friend. I felt as alienated as Johnny did!

"You want to know how *I'd* have ended it? I'll tell you. I saw this line from Thoreau, in his *Journals:* 'Simply to see a distant horizon through a clear air—the fine outline of a distant hill or a blue mountain top through some new vista—this is wealth enough for one afternoon.' For one life! I had it all figured out, how Johnny saw that on a calendar, the way I did, and it grabbed him, and he was converted, that's right. He went to the bank, took out a loan, returned the money, and saw in his mind what Thoreau did. He wasn't whistling in the dark; he was musing and smiling, maybe at the enormous contours of a chestnut tree in full bloom: a thing of beauty he'd failed to see for years, because he'd taken it for granted."

But another student in the class objected strenuously to Fred's argument. Holly saw the story as a spiritual one, with mystery as its essence. For her Johnny was pulled to thievery not out of real, immediate need, but as a means of confronting the dreary, demonic side of his Shady Hill life, a side he all along knew to exist but had to confront directly. The act of stealing was "existential"—a decision to give ongoing evil an embodiment. The story for Holly was an evocation of a pilgrimage, and its ending in no way diminished its ethical thrust: "If you think of grace as God's, then anything, even the most boring or stupid event, can be an instrument of His. What else is there for us in this life but more of the same? We change when we see 'more of the same' in a different light. Johnny's life took on a new sweetness; he was the existentialist hero, a suburban, stockbroker version. The European writers gave us all those murderers and revolutionaries and philosophers as existentialist heroes. Cheever gives us a Connecticut (or New Jersey) commuter, a Wall Street drone, who has a good pedigree, and who seems 'well adjusted.' He becomes maladjusted, and that's part of his salvation. You can't expect the social life to change, or the business life; it's what happens *inside* him that matters. The telephone calls [from Johnny's old colleague, for instance] will keep coming—but they're heard differently."

Such a story begins to work with students because its details remind them of what they themselves have experienced, or maybe what they hope to experience. (Affluent boredom!) Finally, however, it is a storyteller's lyrical magic that insinuates itself into the separate lives of the various readers, eliciting from each one a response that tells as much about him or her as it does about Cheever's fictional creation. Whether "The Housebreaker of Shady Hill" is regarded as a "dark night of the soul," as a bond salesman's epiphany, as a tale of estrangement from and eventual return to a community, as a sardonic commentary on the supposed bliss of America's upper bourgeoisie, or, as one student put it, "an extended nightmare, like we all have sometimes," will depend upon what our own lives bring to the story, as Holly indicated in her commentary: "A story can be turned into an intellectual exercise, but I hope we don't stop there, at least in this class! Cheever knows the world I grew up in; I come from Darien. But I didn't know that world as well as I do now after reading these stories. My experience with that world was the same as Johnny's with his life! Most of it [his life] was spent not paying attention to lots of meanness and dishonesty. When my dad had a coronary and I visited him in the hospital, I never expected to hear him say what he did. I remember every word: 'I should have cleared out of here a long time ago.' I couldn't understand what he meant, at first. He saw the confusion on my face. I thought he might be having mental troubles: you know, a reaction to a severe sickness. Actually, I thought he might be talking about his marriage: my mother and he have had some big brawls in their day.

"But he meant something else. He explained to me what he meant. He told me he wasn't 'born to be a vice president of a paper company.' I knew he'd lost out for CEO, but I thought that had happened a long time ago and he'd gotten over it. But I was way off track, and he saw on my face that I was. He gave me a speech. He told me he'd always dreamed of designing houses and building them—not only being an architect, but a builder. I knew he loved carpentry; it was his big hobby. But I never realized what it meant to him, working with lumber. I guess he thought he'd never make much of a living being an architect and a builder during the Depression and the Second World War, when he began thinking of a career. Also, he came from a poor family, and he thought you have to have the right connections to be a successful architect. Besides, my grandfather had been an electrician, and he told Dad to stay out of the building trades. I think he wanted Dad to be a lawyer, but Dad had no stomach for arguing rights and wrongs in court. He wanted to make money, and he had a college friend whose father was a businessman, and one thing led to another, and he ended up where he is, and he worked his way up.

"He told me that day in the hospital that his life was 'a big blur'; that he never liked his work but kept reminding himself that he'd done what he set out to do, make money and give his family a comfortable life. He told me the worst of it was living where we do. 'It's not any one person, it's everything'—that's what he said. I asked him what he meant. He mentioned Updike and Cheever, their stories. He'd read them in *The New Yorker.* I still didn't know what he was getting at. I grew up there in Fairfield County, and I like it; I wouldn't mind returning there for life. Dad shook his head, and just came up with his laundry list—all the drinking, the exchanging wives, the snobbery, the cutting of people and the fawning all over other people. I tried to argue with him. I said, 'You'll find trouble anywhere, if you stay around and look hard.' But he wasn't going to hear anything that contradicted his opinion. He kept telling me that his heart was 'sick' long before his coronary.

"Then all of a sudden he started crying. When I read in the story that Johnny started crying, I thought of my father crying in the hospital. It was the only time I've ever seen him cry. My mother said *she's* only seen him cry once—when he almost lost his job. She never told us what happened, and my father isn't someone who brings the office home. Every day, when he gets home from the train station, he has those two martinis—and you know what he does while he drinks them? He looks at maps. He's got this 'world gazetteer,' and he looks at states or at countries. He can tell you every state capital. He can tell you where obscure cities in Australia or New Zealand are, or Outer Mongolia! But they practically never travel. My dad does carpentry on his vacation. My mom waits on him hand and foot. I don't think she's had a very happy life.

"I wondered, when I read that story, what Christina was like. What kind of woman did Johnny Hake marry, and why did his mother hate her? There's that paragraph when Cheever describes all the things Christina did: a suburban housewife's full day. But he never gets inside her head or has Johnny try to. We know they fight and are on the brink of separation. But Christina seems always a 'reactor'—that's a word one of the [Harvard Business School] professors is always using. He has his 'personality types,' and one of them is a 'reactor,' meaning someone who responds to what others do or say. My hunch is that Christina would tell us a quite different story about Shady Hill and Johnny Hake—or maybe she'd never think of talking about herself. That's how my mother has managed to keep going. She *never* opens up—not to anyone, even my sister, and they're close. Mom can stand her ground, or argue and argue, but she won't let you inside her heart. Once I asked my sister why she doesn't push Mom to talk about what she's feeling. 'Listen,' she said, 'be glad she's quiet, because if she started talking that way she might never stop!' Maybe that's what would happen to Christina. Who knows? Only Cheever!"

· · ·

The Cheever story "The Sorrows of Gin," a devastating account of another upper-class suburban home's maids and their drinking, not to mention the drinking of their bosses, revolves around a child's moral sensibility. Amy is under ten, and in certain ways over one hundred. She is trotted out for guests and then sent back to mind her own business while her parents wine and dine. Those parents are sadly uninterested in knowing her—and indeed in knowing themselves. They have embraced the forms and routines of their rich, stuffy, pretentious life, leaving their daughter bewildered and scared, a constant onlooker. Amy tries hard to make sense of life—of her parents' false, self-serving morality, and of the hopes and fears she learns to recognize in the people who work for them. In a sense, that is what the story is about: a child's superior powers of observation and empathy. Amy's loneliness does not in itself account for her interest in those whose job it is to care for her. She has taken careful stock of her parents and sees all too clearly their shallowness, their hypocrisy. We have been told it is the other way around—that children are incapable of the broader kinds of human understanding many adults achieve, that the young are wrapped up in themselves. Not Amy Lawton, whose wry, distrustful looks and intent gestures of interest and concern are rationed out with an astonishing shrewdness. She knows all too well the polished emptiness, the dreary desolation, of the dinner-party lives her parents lead. She knows who pays the costs—herself, of course, and legions of baby sitters, maids, and cooks, all doing the bidding of people with the elegant names Cheever deliberately chooses, the Beardens, the Farquasons, the Parminters. The drinking one baby sitter does when off duty (she comes back to work drunk) is a mere footnote to the drinking that takes place in the households of the Lawtons and their neighbors. When Amy gets the idea of emptying gin bottles at home, causing one of her sitters to be regarded as a serious sneak drinker, she is indeed giving expression to "the sorrows of gin": *her* sorrow. She is treated, day after day, as a nice possession, to be enjoyed at times, but put aside firmly for other preoccupations. Amy realizes she can destroy the reputation of an adult by emptying one bottle of wicked magic; she is far less sanguine about her long-term prospects with her parents, especially her father, whose moral piety with respect to the drinking habits of those he employs is not at all balanced by any discernible disposition to look inward and take stock of his own manner of being at work, at home.

This is Cheever's central theme (as it is Fitzgerald's in *The Great Gatsby*): those who are busy climbing ladders, or digging in their heels on the top rungs of those ladders, leave it to others to pick up afterward—to the maids, for instance, in "The Bus to St. James," who pick up the peanuts from the rug. Cheever has a fine eye for the small moments of tyranny or smugness which seriously mar our lives. He is a master at portraying hypocrisy, as in the case of a mother who, when her daughter disappears, seeks comfort in reading the Bible—a Gideon Bible she had once taken from a hotel room. Cheever wants the apparently inconsequential details of life to sound a moral alarm. Even when we meet a woman who protests a school's exclusionary policies and protects her maids from her nasty husband, we are not allowed the comfort of moral relaxation: the woman can embrace adultery without a second thought, and not necessarily as a grand passion, but rather, it seems, in response to a moment's "need."

When students tell me that Cheever is "seductive," I think they mean that he has touched something in their lives. He prompts smiles in us, but we begin to squirm: maybe we are laughing at ourselves, thinking of all those children of ours who get lost, who are ignored, who witness us boozing it up, see us betraying one another sexually, gossiping endlessly about each other, sizing up our so-called friends with envy or discussing them with bragging triumph. "I read Cheever," one student told me, "as if I'm on one of those suburban trains of his, headed home to Westchester or Connecticut, or any pretty, woodsy American township, and I'm just sitting there, in the plushest car, having a drink and looking forward to a swimming party, and feeling glad inside, because out of the corner of my eye I can see the [*Wall Street*] *Journal,* and by God, the market has been up for a month—but all of a sudden the train stops, and there's Cheever, standing outside the train, in front of it, in the middle of the tracks, maybe telling us that we're in trouble, because the train isn't going where we thought, and it might plunge off the track any minute, and we'll end up in hell. Is that a momentary 'bad thought'? Should I shrug it off and have another drink? Or have I been given the chance of a lifetime: to change trains, change my destination?"

Novels and stories are renderings of life; they can not only keep us company, but admonish us, point us in new directions, or give us the courage to stay a given course. They can offer us kinsmen, kinswomen, comrades, advisers—offer us other eyes through which we might see, other ears with which we might make soundings. Every medical student, law student, or business school student, every man or woman studying at a graduate school of education or learning to be an architect, will all too quickly be beyond schooling, will be out there making a living and, too, just plain living—that is, trying to find and offer to others the affection and love that give purpose to our time spent here. No wonder, then, a Dr. Lydgate or a Dick Diver can be cautionary figures to us, especially to us doctors, can be spiritual companions, can be persons, however "imaginary" in nature, who give us pause and help us in the private moments when we try to find our bearings. No wonder, too, Jack Burden in Robert Penn Warren's *All the King's Men* speaks to all of us who have tried to understand the nature of politics, the possibilities, the terrible temptations.

When students studying medicine get to know Dr. Lydgate, when a journalist thinks of Jack Burden and his relationship to Willie Stark, when those who plan to enter government service read *Little Dorrit* or *Anna Karenina* and think of the portraits of bureaucrats in those novels, when a law student meets the lawyers in Dickens' *Bleak House* or *Great Expectations* or *A Tale of Two Cities,* when a university professor contemplates Jude Fawley's view (Hardy's view) of Oxford, when a business school student who has already been in banking or at a brokerage house or in a corporation reads Saul Bellow's *Seize the Day* and meets Tommy Wilhelm, a commodities speculator who can't be as ambitious and grasping as his doctor father and suffers accordingly—there can be a moment of recognition, of serious pause, of tough self-scrutiny.

A former student, Gerard, described that kind of self-searching to me five years after graduation from medical school, his residency training in surgery by then completed: "I'm not ashamed to say that I can talk to that Dr. Lydgate. He's someone I know. I've forgotten all the 'imagery' in *Middlemarch,* and I don't know anymore the English history [Eliot] worked into the novel, and some of the minor characters have disappeared from my head. It's not social respect I want; it's the end of this long, long apprenticeship. I'm finally through, but I don't know quite what to do. Dr. Lydgate warns me I might become a society doc, and with my debts, why not? But I can't quite go down that road, as tempting as it is. There's my dad's voice: he was an idealist to the end, a bitter one, but an idealist. He thought we should sacrifice ourselves for others—and he sure did for us four kids. He worked all day and half the night. He was a druggist; but he was, really, a slave. He opened that store seven days a week. He never took a vacation. My mother worked there, too. They both were slaves—slaves to us kids, to our future.

"I could write my own novel! I suppose everyone could, if they'd only stop and think that way, the way novelists do. You can be too dedicated, or too smart, too clever, or too *anything*, and it'll come home to haunt you. I'm afraid that I'm becoming a workaholic, like my dad; and he's always telling me to slow down and find more time to relax, like he didn't do! He also tells me that money isn't everything—and it wasn't for him. He made money to give it to us; but he did grind away his life to make that money, and I sometimes wonder whether I won't imitate him. I try to think about a novel no one will ever write, with my folks in it, and me, and my brothers and my sister. What would George Eliot do with our family? She'd show the changes in me—how I've stopped telling myself that there are lots more important things in life than making a mountain of money, or even becoming the world's absolutely greatest surgeon. I mean my family, of course. But my wife and I are lucky if we have two or three meals together all week. I never see our little kid. I'm up and out before he wakes up and home long after he's asleep. He's two years old and he thinks I'm a nice stranger who isn't around much, and even when that stranger *is* around, the phone rings and he's always ready to leave. No wonder I'll buy a Megabucks ticket every once in a while! I'd like to become rich and then do some kind of surgical research, with my hours my own to determine. There are days when I think of George Eliot and her Lydgate, and I come to the conclusion that lots of us doctors fool ourselves very easily, and that's what *Middlemarch* has to say to me now, just as it did back then [when he read it in college, and a second time in medical school]. But maybe there's hope. I remember the end of the novel, when she [Eliot] pointed out that you never do know how a life will turn out. (I'm sure paraphrasing!) Well, maybe my friend Lydgate will help me turn the corner—go after what I think is right for me to do, for the sake of my wife and son, and for my own sake, too. I'd hate to end up a driven, driven 'success,' who is bored by what he does, but is always postponing any moral confrontation with himself!"

Gerard and others have learned to regard the characters in novels as persisting voices—friendly and reassuring or sternly critical, depending on the occasion. The irony of Eliot's question, asked in the last chapter of *Middlemarch* ("Who can quit young lives after being long in company with them, and not desire to know what befell them in their after-years?"), is that her philosophical inquiry was not abstract or solipsistic. The destinies of her characters may have surprised her (while she wrote) as much as they surprised any of us who came after her. Her moral investigation has the only permanence that matters—to be lodged in the memory of men and women, even to be taken into account when life seems quiet and comfortable, perhaps in the knowledge that soon a more urgent pace will take over. When a college junior told us in a seminar that she keeps trying to imagine what Dorothea Brooke looks like and what she looked like for George Eliot, she was reminding us, as have others, of Dorothea Brooke's never-ending significance. She is as various as the shapes and contours of our minds. She persists in the thoughts and reveries of all of us who have met her and hold dear her words and deeds, her very life.

· 7 ·

Looking Back

In tahiti, in 1897, at the end of his own life, the introspective painter Paul Gauguin offered the world his "spiritual testament," a thirteen-foot canvas, which he would eventually title, in French, "Where Do We Come From? What Are We? Where Are We Going?" His intention in old age was to look back in hopes of glimpsing life as we live it from infancy to our last years. In a letter he described his efforts: "It is a canvas four meters fifty in width by one meter seventy in height. The two upper corners are chrome yellow, with the inscription on the left and my signature on the right, like a fresco which is appliquéd upon a golden wall and damaged at the corners. To the right at the lower end, a sleeping child and three crouching women. Two figures dressed in purple confide their thoughts to one another. An enormous crouching figure, out of all proportion, and intentionally so, raises its arm and stares in wonderment upon these two, who dare to think of our destiny. A figure in the center is picking fruit. Two cats near a child. A white goat. An idol, with its arms mysteriously raised in a sort of rhythm, seems to indicate the Beyond. A crouching figure seems to listen to the idol. Then lastly, an old woman, nearing death, appears to accept everything, to resign herself to her thoughts. She completes the legend. At her feet a strange white bird, holding a lizard in its claws, represents the futility of vain words. All this is on the bank of a river in the woods. In the background the ocean, then the mountains of a neighboring island. Despite the changes of tone, the coloring of landscape is constant, blue and veronese green. The naked figures stand out in orange . . . So I have finished a philosophical work on a theme comparable to that of the Gospel."

Gauguin's central figure in the picture plucks an apple from the tree, reminding us of the Garden of Eden. His elderly, mummified woman sits huddled in a corner, her hands covering her ears, as if she need hear nothing from outside. The viewer is reminded, perhaps, that in the end we have only what is inside ourselves to contemplate; each of us has a story that contains our answers to the old existentialist questions. I have memories of standing beside my mother as she looked closely at that Gauguin painting—which hangs in Boston's Museum of Fine Arts—and tried to interpret it to me in broad and simple terms. She would remind herself out loud, I recall, that this was the work of a sick man, a dying man, and yet a man, still, of passionate intellectual and artistic intensity.

Tolstoy was another favorite of my mother's; she lived with his stories and eventually died with them very much on her mind. She had read the well-known "Death of Ivan Ilych" as a young woman and had read it several more times before her own moment of death arrived. The story, in fact, was at her bedside during her last illness, and inside the Tolstoy paperback volume she kept a postcard of her beloved Gauguin triptych. Ivan Ilych is a legal functionary, a judge who in middle age falls incurably ill. The story chronicles his demise, tells of his estrangement from members of his family, and they from him, and the progressive, melancholy bitterness and despair that befall this man who knows he will soon be no more, at least on this earth. Of special importance in the story is the manner in which the lonely and frightened patient manages, nevertheless, to draw poignant comfort from his servant, Gerasim, who bears no resemblance to our contemporary culture's psychological-minded "counselors."

Rather, he is a thoughtful, decent person who lacks either emotional guile or self-consciousness. It is precisely those qualities that endear him to Ivan Ilych. Never, apparently, a warm, friendly person, Ivan is now in danger of utter isolation, a death that would precede his death. The end for him seems, in a sense, a continuation of the years that have gone before. It is Gerasim who can lift Ilych's legs up and thereby give him comfort. It is Gerasim who can spark in a forlorn man the human need to be attended. Ilych is not quite Gauguin's mummified elder at the end: Ilych's ears are open to Gerasim's words, and his eyes eagerly seek out Gerasim's presence.

When I was an intern the renowned physicist Enrico Fermi lay mortally ill in the hospital where I worked. I was on his attending doctor's service, and so I had occasion to see him daily. He was a wonderfully sensitive and kindly man, ever considerate of us overworked and insecure house officers. Often he would ask me to sit down and take a piece of candy or fruit from one of the many hospital gifts brought to his room. I well remember the day when I didn't politely sit down and take an apple or a piece of buttercrunch; rather, too tired even for a midafternoon snack, I slumped into the chair. Professor Fermi immediately surmised my condition. I can still hear his words: "Now, doctor, I am the one who is dying! You must find strength. There will be time later for you to sit and think about life." I was embarrassed, startled. I quickly stood up. I went through the motions of checking his intravenous transfusion. I took his blood pressure. I asked him how he was doing. He was, again, affable and candid. He told me he thought he had "a week or two" more. I tried to be reassuring, but he could read my nervous insincerity. He got back to my problems, my exhaustion. He got me talking about my long days and nights of hospital duty, and then he asked me whether I had any time to read. No, I answered with evident irritability and sadness. He volunteered that reading was keeping him alive.

I remember being struck by that way of putting things, on the part of someone riddled with cancer. I asked what he was reading. He nodded with his head toward the table beside his bed, which had been cranked up so that he could sit, rather than be flat on his back. I saw a pile of books and magazines. As I drew closer to see all the titles, he reached for one of the books and asked if I'd ever read any of "them." He was asking about my acquaintance with Tolstoy's stories. Yes, I'd read "The Kreutzer Sonata," and yes, "Master and Man" was one of my mother's favorites. Had I read "The Death of Ivan Ilych"? For a second I was silent. That story, too, was one of my mother's favorites, but I hesitated to say so just then, just there. He had me figured out in a second. He did a quick end run around my awkwardness and condescension: he let me know how grateful he felt to Tolstoy for writing such a story. The expression of thankfulness was warm, earnest, and quite personal—as if, I recall feeling eerily, Tolstoy and he were good friends, and indeed would rather soon be seeing each other. I became apprehensive enough to want to leave the room.

A couple of weeks later Dr. Fermi was dead. I remember talking about his last days on the phone with my father, an engineer who understood the work Fermi and other European scientists had done to help us understand nuclear physics. I also recall telling my father how nervous I was when Fermi had mentioned "The Death of Ivan Ilych." My father was touched and impressed by the stoic Fermi's last days, by the fact that he had been calm and detached enough to be able to understand that death impended and even read a Tolstoy version of what death can be like. I had failed to be as reflective as my father. I had remembered only the anxiety of the suffering Ivan Ilych, his gloom, and had forgotten the last pages of the story—the transfiguration of suffering, the acceptance of death by an ordinary man. For me, then, death had been a constant antagonist; for me, then, Dr. Fermi had been strangely passive before that antagonist.

Later I began to understand a little of what Fermi had been trying to accomplish for himself. I also allowed myself to remember the time when he told me not to worry and not to hurry. On that occasion I had been struggling to get one of the blood transfusions we gave him going again. "Come back later if you have nothing more important to do," he had told me. When I finally got things working, he offered me a drink of juice and told me a story that went back to *his* apprenticeship, when *he* struggled with equipment, and sometimes in vain.

In a sense, that patient (and others far less educated or distinguished) had come upon a truth Tolstoy tries to comprehend in his story of Ivan Ilych's decline. When the dying Ilych asks "What death?" he has begun to realize that the idea of death no longer possesses him. He keeps asking at the end where death is to be found: "And death, where is it?" He repeats the question. Then he finds within himself (not from anyone else) the answer —that "death is finished." The implications of the word "denial," so often used these days, are not what Tolstoy had in mind. Tolstoy was moving his character not through what we call "the stages of dying," but rather toward the revelation that dying has more to do with living than with death. For Ivan, eventually, death and all the foreboding was put aside. I think Enrico Fermi had similarly learned such a distinction, though earlier in his illness than Ilych. The smiling detachment with which Fermi approached his hospital stay, including all its bothersome and oppressive details (plenty of them imposed by my kind), told its own story. Tolstoy makes an important temporal distinction at the end of his story. Ilych's sense that he was "finished" with death took place "in a single instant," but not for others: "For those present his agony continued for another two hours." We may see death at work (and fight it futilely, with mock heroics) in someone for whom the essential matter is no longer death at all. Death is our problem; for the one in the hospital bed, death has already come and gone, regardless of the presence of a pulse, a heartbeat, and a normal electroencephalogram. Ilych's very life, in its coldness, smugness, and self-containment, can be thought of as a living death. As he died he was born—he became for the first time someone who could reach out, connect with others. In that sense "The Death of Ivan Ilych" is about the ending of a deathlike life and the beginning of another kind of life.

Those who are mortally ill often struggle with memories of a past life now endangered; and that struggle is not only psychological in nature but philosophical, moral. When I worked with polio victims during the mid-1950s I was constantly being surprised by the eagerness even children have to use their illness as an occasion for moral reminiscence. A fourteen-year-old youth kept remembering times when he had failed to do his best by his younger sister, who was retarded. He had, as he put it, "walked out on her"; and he hoped that if he got better, he would remember in that future stretch of time what he was remembering then. A much older person (in his forties) was even more zealously self-scrutinizing with respect to the way he had lived his life. He was selfish, he insisted. He had been a tough lawyer and businessman, he wanted me to know. He had given little thought to others, had paid near-exclusive attention to his own ambitions. He kept mentioning, also, the people who worked for him—his secretary, his legal assistants, his accountant, a junior partner in a real estate company. Why hadn't he been more considerate of them? Why hadn't he been more religious? If he had been more interested in God, less in Mammon, might he have been spared this crippling illness? Such questions, put to an overworked doctor in the early evening, are not easily answered. I recall my frustration; I recall, too, a moment of scorn. I think he saw a certain impatience and perhaps coldness in my eyes. I tried to shift our discussion to his medical problems—the paralysis of his legs, his difficulties in bladder control.

Later that night, as I talked with my girlfriend, soon to be my wife, I heard another point of view. Knowing of my love for Tolstoy's stories, my girlfriend kept asking me what *he* might have thought of that patient. I was impatient with her line of discussion—but not unable to counter it in my own psychologically forceful way. Tolstoy, I asserted, had a keen ear for hypocrisy and phoniness and pretense. The patient we'd been discussing, I observed, was not a Tolstoyan figure, eager to repent and redeem himself; rather, he was the same shrewd bargainer he'd always been.

She persisted, however. She asked rhetorically whether anyone who was full of guile and cunning could pass the moral test I was demanding. We discussed, in that regard, "Master and Man," Tolstoy's haunting story of a self-centered, grasping businessman who is not much interested in anyone, his family included, but has rubles constantly on his mind—deals and more deals. On an important moneymaking trip he takes a servant, and though the wintry weather gets worse and worse, he will not stop and wait for the full-scale Russian blizzard to abate. In the end he and his servant and their horse get utterly lost, are at the mercy of relentless winds, terribly cold weather. The driving snow is more than matched by his driving desire to get to his destination, to clinch a commercial agreement at all costs. The horse knows that nature will win out; he stands there, stoically ready for the end. The servant, too, realizes that defeat is inevitable. But not the master; for a while he tries his luck at being bold, ingenious, independent, all the qualities that have helped him to get to the top, to lord it over people on the bottom. He sets out on his own. He plans and calculates, trying to figure out how to get the better of a determined winter storm. Finally he finds himself back where he started—with his servant, with his horse. Yes, he is a master, but he is also a man who at last begins to realize he will shortly die. His servant is still that, an employee who does his bidding; but he, too, is a human being, a man about to lose his life. The story becomes stunningly touching at this point. The master, realizing that there is no hope for himself, sees that the one thing he can do, the one way he can assert himself and his humanity, the one way he can exercise his God-given ability to make decisions and take matters into his own hands, is to try to save his servant's life before he himself dies. He offers the servant his clothing and ultimately himself, his body warmth: he places himself on top of the servant, who is the stronger of the two, the one with the better chance to survive. The servant does survive. The last moments of the master, then, are those in which he struggles with all his much-diminished might to save a fellow human being.

At my girlfriend's suggestion, she and I read the story again. She was trying hard to get me to consider Tolstoy's view of the possibility of redemption, even for the unlikeliest person in the least promising situation, as against what my own viewpoint—a somewhat skeptical psychological reductionism—would allow.

About a week after our discussions began, my patient became even more depressed. He announced that he didn't want to live. He told everyone who worked with him or visited him that he did not want to spend the time left him in a wheelchair. He cried a good deal; he kept asking his Job-like questions: Why me? What did I do to warrant such a turn of fate? He was increasingly regarded as a manipulative egotist by some of us on the ward—a contrast with others, equally ill and paralyzed, who endured their suffering in silence, and contemplated without fanfare their altogether grim prospects. Eventually the nurses and doctors who ran the ward talked of putting him in a special room, to keep him from other polio patients. He was a danger of sorts, we reasoned; he might, with his melancholy self-pity, undermine our ward's morale.

When I went to talk with him about this decision—announce it to him, really—he burst into tears. I had tried to be tactful: we thought he'd feel better away from the other polio patients, some of whom were a bit noisy with their bedside radios. But he was far more discerning than I'd estimated. He spoke with some poignancy about the attachments he was forming to his ward neighbors. At one point he remonstrated: "Don't you see, I've been living in my own rich ghetto world. I'm just starting now to leave it—to appreciate the worth of others who don't have my good luck at making money." He saw, right away, the surprise on my face. He said he believed that we doctors and nurses weren't giving him a break. Yes, I thought to myself, there he goes again—the same self-pitying and self-important character, always "bellyaching." The use of that word—not spoken, but conjured up as I looked at the patient—ought have told me something: I was moralistically on edge, was determined to give this fellow very little credit for any goodness of heart or personal courage.

A half hour or so into that conversation the patient stopped talking. His eyes filled up, but he did not make the sobbing noises he had earlier. I could feel in him a real difference; the silence was more than a mere absence of words. He looked away, out through a window at some men on a platform who were doing some repairs on a nearby building. When I tried to make talk, he said little to nothing, nodding mostly, or shaking his head, or replying with the tersest of phrases. As I got up to leave, he took a pad, wrote on it, tore off the piece of paper with his words on it, put it in an evelope, and handed it to me. I was puzzled as to why he had inserted the message in an envelope while at the same time making quite clear that it was meant for me. I decided he intended me to read it when away from him, and did so. The letter was brief: "Please, doctor, try to find a little charity in your heart for me, even if I'm not a very nice person."

I was devastated. I wanted to go right back and apologize. But I also felt reprimanded, which brought on my irritation and anger. I even noticed that part of me simply wouldn't stop being cynical toward him and pronounced the note to be yet another "ploy." But for heaven's sake, my girlfriend pointed out to me later that day, as we talked about the note and my reaction to it, when was this suffering person ever going to be given any chance at all by me?

We read "Master and Man" together that evening; the next evening, "The Death of Ivan Ilych." I even started rereading *Anna Karenina* after my girlfriend reminded me of Karenin's self-righteousness—a quality, she pointed out, that he had possessed well *before* he'd been wronged by his wife. Tolstoy did not come by his own moral outlook easily; nor was his wisdom, so evident in his storytelling, strong enough to spare him all sorts of personal anguish. His own diaries and his wife's as well bear witness to the sad discrepancy between what he conveyed in his tales and the everyday conduct of his life, with its blemishes, impasses, outright disasters. The more I became immersed yet again in Tolstoy's stories, in Tolstoy's own story, the less unyielding I became to my patient, still on the ward, whom I began to approach with a friendly, even searching, smile, and maybe a bit of shame.

When my patient asked for some magazines, I brought them —and also, one day, a book of Leo Tolstoy's stories, including the two my future wife and I had spent so much time discussing. She thought he might enjoy reading them. I had disagreed. Why expose an already saddened patient to such an array of the blues? But, a bit blue myself—reeling, even, from my girlfriend's frank misgivings about my way of seeing things—I decided to share my book with my patient, who by then had become a more significant moral presence in my life than he, and maybe I myself, was able at the time to comprehend.

When I saw him a day later, he told me he had read "The Death of Ivan Ilych." He said nothing of his reaction. I began to realize, with his silence, that I had really set him up, in my mind, for a no-win situation. If he had told me effusively that he liked the story, I would have doubted his sincerity. (I had long ago decided that he could be manipulatively ingratiating.) If he'd been indifferent to Tolstoy or critical of his writing—well, what else did I (so confidently!) expect, given my negative judgment of his character? A day later he was still reading Tolstoy; and by the end of the week he had obviously gone through the entire book. One morning, as I examined him medically, he told me that he was "through" with my book and that I could take it back "anytime." I did so, noting as I walked away that he hadn't really thanked me.

I opened my newly returned book while standing in the hospital elevator. I thumbed the pages as if to reclaim them. I'd figured out weeks earlier how possessive my patient was about the various material possessions he'd accumulated, but failed, at that moment in the elevator, to take a hard look at how very pleased I was to have back my own book. Suddenly, at the front, I found a slip of paper, not (this time) in an envelope. I unfolded it and read: "Dear Doctor, thanks for lending me this book. On my deathbed I'll think of Ilych and of Vasili." (Vasili Andrevich Brekhunov was the master in "Master and Man.")

I put the note in my pocket, keeping it away from the

book—my last-ditch battle to keep my patient's sensibility separate from Tolstoy's. Several times in the hours that followed, I found my hand reaching into my rear pocket, my hand feeling the white, unlined notepaper on which that brief message was written. Each time, I pulled the paper out, held it up to my eyes, read it, read it again. As I put the paper back in the dark confines of my pants pocket over and over, I myself thought of Ivan, dying, and Gerasim, attending him with such eager, unquestioning attention; and of Vasili and his great dreams of wealth, the servant, Nikita, and that lively, responsive horse, Dapple, whose dignity required no words of affirmation; and of Lev Tolstoy and Sonya Tolstoy, the two of them so devoted to each other, some of the time, and yet often so at odds—each decent and openhearted and principled, and each capable of being narrow, spiteful, vain, smug, and, not least, petty and vengeful toward the other.

Hadn't the time come, I asked myself at last, for me to give this one "difficult" patient the benefit of the doubt, and thus spare myself any further acts of callousness, any further self-inflicted humiliation? The next day, as I was preparing to leave the ward, I went over to see him. Now *he* was reading *Anna Karenina;* his wife had brought him their entire Tolstoy library, all unread, she had told me, thereby giving me one more opportunity for stupid condescension. He looked up; I gave him a note I'd written earlier that day. He did not read it while I was there. I realized even as I handed him the note, folded, that I didn't want him to read it in front of me. I beat a hasty exit as he sat there holding the piece of paper in his hands. "Please," I'd said in the note, "know that I admire your courage as you take on the rehabilitation [which had just begun in earnest], and if I can be of any help, anytime, let me know." Then I'd added (our bond): "I love Tolstoy, and I love seeing you read his stories."

So began, over the many weeks of his recovery, a friendship as we discussed, briefly, the Tolstoyan oeuvre; book by book he made it his own. When at last he was getting ready to be discharged, on crutches, I felt downhearted. I wanted him to go home to his wife and children, but I also knew how much I'd miss him, miss those talks. I had often saved them for the end of a tiring day, when a few words between us more than made up for the glass of bourbon to which I would otherwise have paid a visit after work. On our last meeting in the ward, he had a present for me, nicely wrapped, which I knew he did not want me to open in his presence. I had brought something for him. We exchanged presents, and as I watched him hobble a bit awkwardly toward the elevator, my hands held on self-consciously to his package—a book, obviously. Minutes later I opened it: a biography of Leo Tolstoy. I had to make sure I was smiling, appreciative, but not visibly upset: some nurses and doctors were nearby. Thank God for the hay fever season.

. . .

Not all patients get entangled in Tolstoy; nor do all their doctors. Tolstoy himself, as he got older and looked back at his life (in *Confession,* for example), was tempted to turn away from the intellect, including those precious narratives he had wrought: "Many times I have envied the peasants for their illiteracy and their lack of education." He was not being gratuitously anti-intellectual. He knew, surely, the risks of an avowed humility—knew *its* pride. Tolstoy was referring to the obstacles both to religious faith and to a reliable trust of one's fellow human beings. He wondered consequently (at least sometimes) whether the intellect wasn't more a hindrance than an aid to human relatedness, to our moral life. But he knew better, at other times, than to take the risk of celebrating ignorance per se. He sought not "illiteracy" but rather a kind of innocence, though he also knew that few (if any) of us can find such a quality of mind and heart for ourselves. Still, like other writers and moralists who preceded him and followed him, the special dangers of intellectuality preoccupied him even as the humble of this world had a constant appeal, not unlike the kind Faulkner, say, felt for the black people of the South, of Yoknapatawpha County.

Young people in colleges and graduate schools, not yet Tolstoys and Faulkners, wonder, as they picture themselves in the future living reasonably privileged lives, what their own values should be. Will they themselves be thoughtful of others, respectful of them, regardless of their background and education? One student, named Arthur, wanted to discuss his college teachers—dryly at first, then with an alarming acerbity: "We are graded, you know, on our 'moral reasoning.' I've taken two courses with that as the title. Each time I got an A. Not bad! I guess I'm certified as a pretty moral person—or am I? I mean, am I a smart moral thinker? Am I moral in the way I act during the day and at night? Are the two the same? We never even got into all that. If you push questions like that with some of you guys [professors], you get told: hey, *this is a university!* That means, we're here to *think*—as though talking like that isn't doing more than thinking! The other day we came to class and the door was locked. We stood outside waiting, and then the professor came. He wanted that door opened. I don't blame him. He went looking for someone to open it. He couldn't find anyone. You should have heard him barking at some poor museum guard; you would have thought that guy was working for Satan. My girlfriend was embarrassed. Why embarrassed? [I had asked.] Because we were there with him [the professor], and we weren't telling him to shut up. Finally, someone came and opened the door. The man with the keys sure didn't get a thank you; he got a dirty look. Then we all filed in and heard a lecture on 'moral development.'

"In a sociology course we study racism, and we read all these books about ghettos. This university is one, too. It's even got walls around it, like the ghettos of Europe. My girlfriend and I were watching some of the professors going into the faculty club, and coming out, and we decided that they knew a lot about atoms and molecules or ancient Mesopotamian culture. But when a guy gets up and lectures on morality, you expect him to practice what he preaches. That's an old saying, and it ought to be branded on all of our arms or hands so we'd see it every day. I'm sure we'd all forget that it was there, though! I feel sorry for the university employees who don't take courses in 'moral analysis'—but they have to be well behaved all day long. If they slip up, they lose their jobs! They serve us food, and all we do is grouch and groan and say we *can't stand* what they have for supper, and meanwhile those folks are standing patiently behind the counter, serving us, and hearing us talk like we do, and if they get one 'please' and one 'thank you' per half an hour, they're lucky.

"It's true, I'm being unfair. Lots of us *are* polite. But there are plenty of snot-faces around, and I'm willing to admit being one of them plenty of times. You get all caught up in yourself, and you need to be pushed out of yourself, and it's hard, when everyone's into himself around here, and that's what gets rewarded, your achievements, what you do. We must look like some strange breed of animal to the people who work here."

He was, I thought, more than a bit wrong-headed. He emphasized too much the negatives, the solipsistic arrogance that, unquestionably, he observed among both professors and his fellow students—failing, meanwhile, to acknowledge the sensitivity and thoughtfulness of plenty of undergraduates, and of their teachers as well. His interest in those who worked at the college was, however, relatively rare. He himself came from a quite wealthy home, and for reasons of his own—the home was also quite troubled—he had found a good deal of solace in the company of maids and handymen, especially one jack-of-all-trades who cut the grass, planted flowers and vegetables, pruned trees, cared for the swimming pool, lugged lawn furniture around, raked leaves, and, as the student once said in summary, "did everything." One of the things that middle-aged worker did was to be a reliable and kindly pal to a growing boy whose own father was quick-tempered and decidedly alcoholic. Out of such a home life came Arthur's persisting interest in the lives of working-class families, and also a certain tough skepticism with regard to the manners and assumptions of the privileged. Still, at that time, as I heard him praise Tillie Olsen's stories and rail against the college he attended, I thought he made important points, ones not always made by others in the community, but overstated his case. Ironically, he fell victim to his own kind of smugness—a hazard for any self-appointed critic. We ought be apprehensive lest we share in what we condemn, lest the act of condemning bring us close to what we may have properly singled out for blame.

As for the Olsen stories, they most certainly prompted an outpouring not only from Arthur but from many of his classmates. In fact, as mentioned earlier, I have used the four stories in *Tell Me a Riddle* with students of all kinds. "I Stand Here Ironing" is a story of wide and suggestive appeal among educated young people because it reminds them of the past as well as the future. The story centers on the cultural and social distance between the therapist-listener and the poor, working mother, who can readily be regarded as a stand-in for the world's patients or clients or customers. But the story has another side: the look backward of a woman now in middle age. Young people use the story not so much to warn themselves against a future complacency as to recall some of their memories of times when they felt misunderstood and were unable to give full expression to the intense frustration of that misunderstanding. Even as the speaker tries to explain to herself (and, presumably, a therapist) how she and her children came to be the people they now are, a student reading the story can entertain a similar desire to set the record straight. College students have heard themselves called all sorts of psychological or sociological names. No wonder so many of them, in turn, try to comprehend themselves, as does the woman who stands ironing.

Each of Olsen's stories takes on the conventional world. The world in "Hey Sailor, What Ship?" is one that readily scorns a drunk, yet accepts and even relies upon his generosity. Whitey is a hurt idealist. He, too, looks back—to the days when life was an adventure, when the body was vigorous, when a sense of timelessness prevailed, when hope lay around every corner. Now his former friends argue over him, and their discussions are profoundly moral. Ought one remain loyal to a person, no matter how serious his troubles? Does a day come when the past ceases to be important—when a person has a right to walk away, to call someone a former friend? What, indeed, are our responsibilities to one another as acquaintances or friends, and for what reasons ought such obligations be set aside?

The ultimate "friendship," the union of two human beings grounded in law, in social custom, and, for many, in religious sanction, is marriage. It is a marriage of forty-seven years that Tillie Olsen examines in the title story of her collection, "Tell Me a Riddle," as poignant and disturbing a retrospection as one can find in contemporary literature. Frequently, when discussing this story, I find myself mentioning Shakespeare's *King Lear.* As Shakespeare and Olsen remind us, a father can fail wretchedly to have any real moral or psychological comprehension of his daughters, and a husband and wife who have been married almost half a century can be virtual strangers to each other, their memories stimuli to a mutual estrangement. Olsen makes clear that her elderly man and woman are a puzzle not only to each other but to their children and grandchildren as well; and she especially emphasizes the husband's complacent, self-indulgent habits of mind—his essential egotism, his endless interest in waking his wife to initiatives and postures that serve his interests alone. They look as though they have survived a long, solid marriage; in truth, they have survived only each other, or so the wife especially feels. The husband's evident, outspoken shallowness is, perhaps, his own way of being confessional—as if he has found nothing in life to help him become more complex, more introspective.

For those yet to marry, or for those well into marriage, this story can give the kind of pause a medical student named Joan described one summer as she worked on a project with poor and elderly people in Boston: "I wish at times I'd never read that story! It makes me wonder about some of the people I work with, the ones who almost seem too cheerful, like Tillie Olsen's old man. But what is anyone to do—try to strip away all the pretenses, when the people are in their seventies and eighties? Who am I to do that? Who is anyone? I wonder about the 'message' of that story; I mean, I know it's about a lifetime of resentment welling up in someone who is near death—but after a while I began to feel sorry for the man, and I felt another story could have been written about him with his point of view explored more sympathetically. At the end there's a bit more 'balance'; they're together with less bitterness—well, really, she's near the end, and so hasn't any energy for saying anything. But it's a sad story, and it makes you wonder about your own family and your own future life. My grandparents have been married *over* fifty years, and everyone thinks of them as an ideal, lovely old couple. But I know what's there, under the surface. I've heard my mother talk to my dad, and I've seen and heard for myself. They have terrible arguments, even though they're nearing eighty. But when they take a walk or enter a plane to travel, everyone thinks: a beautiful old couple!

"I don't blame them for pretending to be what they're not. If anything, it's *us,* what we do: we spin our illusions. My mother says she built her whole life around *not* having a marriage like her parents'. She and my father *never* fight. They both get depressed, but they don't fight, at least not in the conventional way. They fight, if you ask me, by avoiding each other for a while. Then they get back together. I can tell the state of their marriage by the chairs they use in the living room!

"I hear a lot of memories at work, when people talk with me. I'm really a stranger, even if I've been there a month, but they see me every day. They tell me all the wonderful moments of their youth, their early married life, and I smile and I'm impressed. I hear them tell one another the same stories. Anyone new, of course, gets the full treatment, laughs and tears and all. I'm not being cynical; I'm just being as bluntly realistic as I think I should be. I've got my life ahead of me. If I fool myself about what's going on around me, I'll end up being a performer—and that's a lot of what I see those people doing all day: they perform. They say things and do things that get you to smile at them and admire them and think to yourself, hey, they're saintly folks, and just a step from heaven. Actually, some of them are as mean and envious and sour in disposition and moody as we young folks are.

"I think Tillie Olsen is stripping something from our minds—the romantic Norman Rockwell idea of the old couple, respected and loved by everyone, sitting there, holding hands and sharing their memories with anyone who will listen. If that's what being old is, putting on a song and dance like that, Tillie Olsen and her old woman don't want any part of it. Olsen's story is tough—and the lives of that couple were tough. They struggled, and their marriage paid a price; and when she was sick, she was defiant, mostly, and he tried to placate her, win her over, and finally he defended himself, and in a way he tried to defend a certain integrity he felt the marriage possessed. But much of the story is a reminder of what can happen in anyone's life: you get into a groove with another person, and you're both stuck in it, and it's how you act with each other, and you can be completely blind to what's really going on. All you try to do is keep what's familiar perking along.

"I've been recalling some of my own life—how I went steady for two years with a guy I never really got to know and, I guess, never wanted to know. Why? The answer is hard for me to admit, even to myself, never mind to say to someone else! There are times when you're scared—scared of having no one, of being considered a very young version of an old maid. There are times when you hold on to anything—rather than someone!—if you know what I mean. You're being dishonest with the other person; you're probably being dishonest with yourself. It can become a habit, too. It can be the basis of a marriage—even of a marriage that lasts. That's what Tillie Olsen is saying.

"My boyfriend—he was good-looking and charming, and we could go places where it would have been awkward for either of us to go alone, and we enjoyed being together. What I really mean—well, we liked the *idea* of the two of us, and we liked what people did with us: signaled their pleasure in being with a 'nice young couple.' That's what my grandparents called us. That's what people say about *them,* and they eat it up: 'A nice old couple!' When will we learn to be more honest with ourselves, more honest with others? That's not only a psychological question; it has to be, at some point, a moral question. That's why I see the Olsen story as a moral tale, more than anything else."

A "moral tale" gets a moral response from Joan, whose work,

fortunately, enables her to be especially touched by a book. Obviously, even a novel or story that produces moral implications for some readers can fail to have such an influence on others. In my experience with college and graduate school students, a novel can exert greater moral impact when there is a shared personal experience a class can summon in its discussions. More of the students respond to the fictional world and its moral quandaries, and their response is deeper, more strongly felt, more lasting. When medical students, for instance, have been working in a ghetto clinic or with alcoholics, elderly people, or the homeless, what they bring to a weekly seminar discussion of Tillie Olsen's stories can be illuminating indeed, and sometimes quite poignant.

Students have referred to a "double whammy" effect—a notable heightening of interest and reflection that occurs when they work with alcoholics in a soup kitchen or treatment center or outreach program and then sit down in a time of leisure to read "Hey Sailor, What Ship?" Afterward, when we meet to talk about Whitey, we also discuss the men and women they have met in recent days and, not least, their own reactions. These young people want very much to be of help, yet find themselves confused, irritated, and inclined to doubt their own usefulness, maybe anyone's. Under such circumstances, thought and perception become intimately connected to feelings aroused by past experiences. I do not propose, of course, that we set aside careful analysis, either of a text or of an experience, personal or professional. My emphasis here is simply on the moral force literature can exert, its visionary side.

A medical student of mine named Thomas, who was also an erudite English major with considerable literary talent, put the case forcefully during a seminar for medical students who had worked among Boston's medically indigent: "I'd read all those stories before. It's one thing to read Tillie Olsen or Ralph Ellison or William Carlos Williams in a literature course; it's another to read them now. It's different. We're in the situation Williams describes in those 'doctor stories': worried or nervous or disappointed or frustrated or irritated or angry as hell. When you've experienced those emotions in the same way a writer has described them, and you're reading his descriptions and then discussing them with other people who are in the same boat you're in—well, that's *something.* I've never thought of stories or a novel as a help in figuring out how to get through a working day. Sure, you read a novel like *Magic Mountain* or *Middlemarch* or *Arrowsmith* and you ask yourself what this life means, or how should a doctor live his life. But those are 'big' questions, philosophical ones. When you're learning to be a doctor, you probably read those novels a little differently than you did when you were in college. But now, when you're coming back from the work we do in the clinic and you pick up some of those stories, you feel as if the author knows you personally and is talking right to you. You can't forget the stories, and you think of them not only here, when we're talking about them, but in your car, or when you're walking, or when you're out there doing your work. I'll see someone or I'll hear something, and 'click,' I'm back there with a story, and thinking about what we discussed, and what the novelist was pushing us to consider. Sometimes, the story I read merges with the situation I'm in, and I'll also think of what we all said in class, and then I'll have some [personal] memory come to mind, and I can't let go of it all, and it all becomes part of me, I mean, it all affects my attitude toward the people I'm with, in the clinic."

Thomas does not pretend to be a precise psychological theorist or a penetrating literary critic. Rather, he wants to connect an intellectual experience with an aspect of his everyday working life, and to link both with the longer haul of his life—with the memories that a story or a work experience can sometimes bring to the surface. When he mentioned such memories I had one myself—one of my college teachers discussing Dostoevsky's novels. As he lectured on *The Brothers Karamazov,* he read aloud his favorite passages. In my mind I was comparing his favorites with mine—and with my mother's, my father's. I well remember him reading toward the end of a lecture hour from one of the last paragraphs of that long novel, a portion of Alyosha's address to the young people who so revered him. I didn't know the passage by heart, but I did know it well enough to recognize it immediately, and during that day it would be the passage that my mind kept finding itself remembering. Back in my room that night, I looked it up and read it again—as I did many years later when Thomas prompted Alyosha's words once more: "My dear children . . . you must know that there is nothing higher and stronger and more wholesome and good for life in the future than some good memory, especially a memory of childhood, of home. People talk to you a great deal about your education, but some good, sacred memory, preserved from childhood, is perhaps the best education. If one carries many such memories into life, one is safe to the end of one's days, and if one has only one good memory left in one's heart, even that may be the means of saving us."

Dostoevsky, in a rhetorical moment, is trying to move us all past rhetoric, including the significant recourse to it in what gets called "education." A memory is, of course, a story, an aspect of experience that lives in a particular mind. These days, given the emphasis on Freud's work (and *its* existence as a cultural memory), such an admonition may seem unnecessary. Yet Dostoevsky is not advocating memory as a means of clarifying our mental problems or as a distorted representation of what "really" took place. He is talking instead about what we might consider memory's more "superficial" sense, a recollected moment in which someone has tasted of life, a moment forceful enough, charged enough, to survive many other moments. Perhaps Dostoevsky is suggesting that an especially vivid memory itself survives as a monument of factuality (whether or not the memory is of a real event)—and helps *us* survive. Without such compelling memories, we are not ourselves, but rather anyone. A memory is an event endowed with the subjectivity of our imaginative life. Dostoevsky's "good memories" do not include nightmares or moments of horror; they are not per se sufficient food for the moral imagination. If we are trying to find our bearings, all the wrong directions we recall, all the dead-end streets we have traveled, won't in and of themselves be enough.

Even as Dostoevsky urged us to look backward for a saving memory or two, and even as Tillie Olsen's couple look so relentlessly and searchingly backward in "Tell Me a Riddle" (as do Tolstoy's Ivan Ilych and his "master" in "Master and Man"), the last minutes of a university class, and especially the final papers it generates, can be an occasion for complex recall of all kinds. Teachers have known for years that passages such as the one I have just quoted from Dostoevsky lend themselves well to personal reminiscence. I remember the last meeting of one seminar, in which a college student named Richard who had read Hemingway's *The Old Man and the Sea,* as well as Tolstoy's story of Ivan Ilych's death, talked about his own "old age," the phrase he used even though he was twenty: "I'm young if you look at my birth certificate, or if you look at me walking down the street. But I've got a good chance of dying in a few years. If you could look at x-rays of me, that's what you'd say. A few years ago I didn't know what a radiologist was; now they are my shrinks. They look inside me and tell me what my prospects are! They say, 'It's getting smaller,' or 'It's about the same.' I took a course in psychoanalytic psychology, and the professor told us that the id means the 'it,' the unconscious. For me, since I got sick, the word 'it' has meant only one thing: the tumor growing inside me. It's like the unconscious! You can't see 'it,' and you think your life is yours, to control through your actions, but there's something silent and powerful deep within you that's got a big say, maybe the biggest say, in what's going to happen to you.

"I feel I'm an old man; I might be dead within two or three years; or I might survive for much longer, ten years; or I might even be cured, especially if research brings new 'treatment modalities,' an expression I've learned since I got this lymphoma, and I'll tell you, that's an expression that has as much meaning for me as 'Father, Son, and Holy Spirit'! You feel a lump under your arm one day, and you think it's a bug bite. I'm allergic to bees, so why not! The lump won't go away. It will, eventually. Then you have to get a medical exam, because you're an athlete, and you go see the doctor, and you actually forget to mention the lump to him. You want to know something: it wasn't 'a lump' back then. It was 'a bite,' or 'a swelling,' maybe a 'mass.' A 'mass'? I'm Catholic: I know what 'a mass' is! Oh, I wanted to laugh back then, and make stupid jokes like I'm making now! I was scared out of my mind. I remembered Hemingway's Nick Adams stories, and his battle scenes, and then that 'old man,' taking on the ocean and its life—all that crossed my mind as that doctor started running down the list of what that 'mass' could be; and I guess I didn't want to pay attention, because I could see in his eyes what he believed, but I didn't want to see in my own eyes what I'd seen in his. I'm not sure, even now, how to describe in words the terror I felt, and my immediate wish to escape by cracking jokes. I did a lot of joking a while ago, and I still do.

"I'd read Hemingway many times, and I'd read some Tolstoy, but it was a different story the second time around, so different I couldn't believe I'd ever read either before. I used to be 'interested' in Hemingway's stories; but when I read them after getting sick, I was into them, so much into them that I'd read a page or so and stop and think. The same with Tolstoy: I'd take a walk and I could see Ivan, and see his servant (I forget his name, but I have a clear picture of him in my mind). I could see Hemingway's 'old man,' fighting for his last victory on that boat. I'd talk with him—not out loud! I began to think of all the novels I'd read over the years, and scenes from some of them came to my mind. I'd read *Jude the Obscure* and written a long paper on it, and I knew a lot of the symbolism Hardy uses; but what I thought about for the first time was—that Jude died young. He was less than thirty, as I recall. No teacher ever pointed that out to us. I'm not saying it's important. It is to me, though, *now.* Hardy doesn't give Jude the kind of dying revelation or moments of awareness that Tolstoy gives Ivan Ilych. Jude is found dead. Hardy puts him to sleep. I kept comparing those two ways of dying. Which kind to want? At different moments I wanted one or the other. I thought of the other ways—dying like Anna Karenina did, just throwing myself in the way of a moving object: not a train but a car, maybe. On the way back to school I ended up sitting near the emergency exit on the plane. I kept staring at those words, 'emergency exit.' I kept looking at the handle. It all seemed so simple; all I need do is turn the handle and leap. In a few seconds I'd be gone. No more of my fretting and fussing; no more of my foul words and my envy of my roommates—for being healthy. I used to have my petty moments with them, when one of them would beat me out in an exam, a grade. Now I *really* have something to envy them for!

"The whole novel [*Jude the Obscure*] is different to me. For me, now, the novel is about death—Jude's gradual loss of all self-respect, all the innocence and decency and thoughtfulness he initially had. The professor who assigned the novel kept saying that it was Hardy's last one. When I was rereading the novel recently I remembered that remark. Jude's death was Hardy's death as a novelist. He'd given us his struggle to confront the universities and the Victorian culture, all the hypocrisy and pretense and snottiness. He'd said all there was to say—so best to die in silence.

"Reading books now is my way of *not* being silent. I had trouble for a long time figuring out what to think about all this, and my dad said I should go talk with someone. I didn't know at first what he meant. Who should I go see? Why? To talk about what? I thought he meant I should see the dean, about planning my schedule—because the radiation treatment, and possible chemotherapy, might 'waste' me, so I'd need time off. But he was riding on another track!

"He knows a psychiatrist—that's what he started pushing on me: go get your 'problem' worked out. He's a lawyer, and he's always calling in those 'expert witnesses.' They all contradict each other, but he still relies on them. Not me! I said I'd go 'talk with someone,' yes—I'd even seek out several people to consult. He looked at me closely. He was surprised and he was curious. I didn't go any further, so he did. Who? I told him some are men and some are women. I waited. He asked again: who? I gave him more information: they came from different countries. He began to look at me as if I really did need a shrink, more than even he had thought! So I came clean: I said they are alive and dead, but they are all really wise folks, and I'm sure he'd approve of them. I didn't smile; I was serious—*dead* serious, you might say! He just didn't get it. He looked at me as if I needed to go right away to the nearest hospital and be put in a whirlpool bath, then given medication. That was when I gave him the names—Hemingway and Tolstoy and Hardy and James Agee and Tillie Olsen and Walker Percy and George Eliot and Dickens and Dostoevsky. I just poured them all out, and he still wasn't sure where I was going, but he looked a little less alarmed. My mother had been listening from the next room, and it was then she came in and started laughing, and told me she was glad I'd appointed a 'board of advisers' for myself, and she thought I'd made some great choices, and *she* was going to 'consult' with them, too. We all dissolved in smiles and Classic Coke with lots of cracked ice!

"I've read [Agee's] *Death in the Family;* I've read a lot of Dostoevsky—*Crime and Punishment* and *The Brothers Karamazov,* both for the second time. I keep reading and keep thinking about what those writers have to tell me about life and death. I think of that old fisherman, trying to get one more big one, and fighting it out with the sharks, pulling on it [the marlin] and taking it in, finally, even though his skiff is half destroyed and the fish mostly eaten by the sharks, and coming into port with it. I wonder whether I'll win my big struggle with 'it'! I'm trying to rope 'it' in; I'm trying to prevail, and I might just not be able to win; but I have no choice but to keep trying. I get preachy with myself, as you can see. I wish I could talk—talk to myself—like a good storyteller: no sentimentality, only emotion that I've earned, that my life has earned. I feel really inadequate, compared to that 'old man' of Hemingway's, or to Jamie [in Walker Percy's *The Last Gentleman*]. Jamie and I are about the same age—two young men who've got cancer. He died of leukemia, I think. 'A powerful death bed-scene.' That's the kind of talk I used to love! Now—well, I think of Jamie, and I picture the sweat on his forehead, and I feel it on my own, and I can see him lying there, stoic and decent, and I hope to God I'll have some of that *silence,* that *acceptance,* in me when the time comes, and it could be sooner than I want to think."

Richard did fall silent, almost determinedly so. He had made the point of points to himself—that knowledge of death as an inevitability for all of us, and knowledge of its psychological consequences for the person who is nearing the end, are not to be confused with the actual experience of dying. I began to realize that Richard was trying to let go of himself, find in fiction another world, which, ironically, helped him explore, by indirection, aspects of his own world—aspects that otherwise (by reason of "fear and trembling") he found to be off limits. He was trying, really, to imagine the unimaginable, with the help of those who are quite good at putting the imagination to work. He was doing so now, moreover, not only for aesthetic reasons but for an urgent moral reason: how ought I bear myself, if at all possible, under these extraordinary circumstances?

Although this bright, able student was pragmatic and even science-oriented in his studies, he was not beyond describing a visionary moment, a breakthrough with respect to the way he regarded both himself and the (threatened) life he was living. Richard found himself at a Boston Symphony Orchestra concert listening to the Mozart *Requiem* and thinking of the brevity of the composer's life. Tears came to his eyes. For a moment he believed he himself was about to die—"right there in Symphony Hall, and with a requiem being played." Mozart had not only consoled him; Mozart had changed him somewhat—offered him a message, not unlike the kind a novelist can offer: a new, revelatory, transforming sense of what life can mean. Months later he would find himself hearing the Mozart piece; envisioning his version of Ivan Ilych or the master of "Master and Man" as they struggled with death; and somehow not feeling anxious or unsettled or fearful, but instead calm, even reassured that Tolstoy and Mozart and other writers and composers were his companions, knew exactly what he was experiencing and what he might unpredictably experience. He repeatedly spoke of certain music, of certain characters in stories, as "all-essential" to his "survival." He had been touched deeply; he had been offered the wisdom of others in such a way that it was truly and unforgettably his.

So it was for Richard, so it has been for many of us—going back, way back, to the earliest of times, when men and women and children looked at one another, at the land, at the sky, at rivers and oceans, at mountains and deserts, at animals and plants, and wondered, as it is in our nature to do: what is all this that I see and hear and find unfolding before me? How shall I comprehend the life that is in me and around me? To do so, stories were constructed—and told, and remembered, and handed down over time, over the generations. Some stories—of persons, of places, of events—were called factual. Some stories were called "imaginative" or "fictional": in them, words were assembled in such a way that readers were treated to a narration of events and introduced to individuals whose words and deeds—well, struck home, or, as some of my students with studied understatement have put it, made an impression that lasts "longer than a few hours." "Longer" for Richard turned out to be longer than he had dared hope possible. Survival did not diminish his interest in the characters he'd met—Jude and the "old man" and Ivan Ilych and Olsen's elderly couple. On the contrary, their presence changed the shape of his life, prompted him in his years as a law student, then as a federal judge's law clerk, then as a member of a prominent Boston law firm, to keep certain texts at his side, stories that helped him as he (in his middle twenties) went through his own story with growing hope.

· · ·

When I began teaching in various parts of the university—the teaching that would eventually give rise to this book—I intended to make an essentially psychological inquiry. I was interested in the ways students respond to literature—there would be different ways, I speculated, for different kinds of students. I reasoned that medical students might well take to *Middlemarch*, say, in one manner, law or business students in another manner. Some novels, after all, must surely address particular audiences with special poignancy or persuasiveness. And I expected that students, who sort themselves out by choice of career, would also sort themselves out as readers: lawyers would be one kind of reader (scrupulous, penetrating, argumentative), doctors another kind (inclined to a bit more empathy, perhaps), teachers interested in the children in a story, businessmen inclined to analyze characters' prospects and compare them with the prospects of people in "real life."

Those were, of course, naive and somewhat absurd generalizations. In truth, I found an astonishing range of responsiveness in my classes—to the point that my stereotypes contributed very little. I discovered, in time, that *All the King's Men* can work as wonderfully with a group of medical students as William Carlos Williams' "doctor stories" can work with students interested in public service, journalism, or law. The decisive matter is how the teacher's imagination engages with the text—a prelude, naturally, to the students' engagement.

Over time I began to learn from the students that constructing a good reading list involves not so much matching student interest with author's subject matter (though there is no reason to ignore the pleasures such a correspondence can offer) as considering the degree of moral engagement a particular text seems able to make with any number of readers. "This novel won't let go of me," a college freshman said to me about *Invisible Man*—and the student was white and from a wealthy, powerful family. Moreover, he was at pains to let me know that he hadn't ever been especially interested in the racial question of the United States and that a reading of the Ellison novel had not at first pushed him in such a direction. What, then, did he make of the novel, and what are we entitled, as his critics or as critics of the novel he has read, to make of his response? In his papers and in his contributions to the class discussions he made clear what Ellison had set in motion—a new awareness of himself as "ignorant" more than "prejudiced." He kept pointing to the irony that for years he had traveled to every continent, including a brief stop in Antarctica, yet never went near Harlem, less than fifty miles from his parents' Connecticut home. Such an irony is, of course, not all that singular in American life. But this student did something more with the Ellison novel: he worked himself into it, connected himself to the central character in such a way that some of Ellison's "invisibility" became for an eminently privileged youth of eighteen a means of self-recognition: "My parents are so damn busy with their lives that they usually want us out of the way, and the longer the better. No wonder they sent me to boarding school when I was only twelve, and no wonder I was off at camp from the age of seven onward. When I was finishing *Invisible Man* I remembered a time long ago; I must have been six or seven, maybe. My mother and father were talking about my brother and me and our older sister. They'd had their Manhattans; that's what they always drank. I heard them say they were going on a vacation, and I heard my dad say something that meant he might worry about us, and then I heard my mother say, 'Out of sight, out of mind.' She wasn't a bad person, but she sure loved a good time. I think we were invisible to her—out of her mind a lot of the time. All those maids took her place. It took that novel to get me remembering!"

A memory can be sprung loose by an encounter with quite another world. That student had found John Cheever's stories all too familiar; they gave him an occasion to yawn and think of Manhattans and sailboats and tennis games. Even "The Sorrows of Gin," with *its* maids, failed to rouse him. But Ellison's novel brought him an insight: he had learned not to notice because he himself had been persistently ignored. With a renewed awareness, he could stop and think not only about America's racial problems but his personal ones as well. The joining of the two in his mind turned out to have consequences not easy to imagine: a real and tenacious concern for ghetto children, an active college life of volunteer teaching in a school attended only by black children. When I heard, one day, that this young man, as a senior, had read *Invisible Man* a second time and even brought it home and asked his parents to read it, I was reminded yet again that a compelling narrative, offering a storyteller's moral imagination vigorously at work, can enable any of us to learn by example, to take to heart what is, really, a gift of grace.

·8·

On Moral Conduct

So OFTEN when I try to grapple with moral matters, my mind goes back to earlier days: my college years, when I worked hard to fathom William Carlos Williams' *Paterson,* and my medical school years, when I visited him fairly often. In the early parts of *Paterson* Williams is quick to scorn the abstract mind at work, whether in the construction of verse, religious ideas, or moral philosophy. He will make a "reply to Greek and Latin with the bare hands." Unlike J. Alfred Prufrock, he will, indeed, be found "daring." He deplores a contentedly aesthetic poetry. "The rest have run out— / after the rabbits"; but he will stay, and provoke "the knowledgeable idiots, the university." He dares assert himself as "sniffing the trees, / just another dog / among a lot of dogs." He quotes from the past, draws from John Addington Symonds' *Studies of the Greek Poets,* but not as so many others have, to applaud the established, the traditional. He will reach for the ordinary, the local, will hope to connect in mind and heart with the energies of this century's ordinary American working men and women.

Williams issued warnings to himself in *Paterson.* They were moral as well as aesthetic. He worried about his "craft" becoming "subverted by thought"; worried that he would end up writing "stale poems." But he was ready to take on more than the aridity of pedants. His repeated call to arms, the well-known phrase "no ideas but in things," is a prelude to distinctions he kept making between poetry and life; between ideas and action; between the abstract and the concrete; between theory and practice; and not least, between art and conduct. As he keeps stressing the importance of testing thought, whether the poet's or the moralist's, by measuring mind against conduct, he keeps submitting himself as an example: the doctor who attends patients regularly and who writes poems such as *Paterson*—and who is, thereby, in double jeopardy, because he may well fail to practice what he preaches to patients and readers alike.

Williams uses the device of a nameless woman correspondent to confront himself, to provide a powerful Augustinian arraignment of his prideful side. These prose segments interrupt the poetry—as if the author is reminding himself and the reader that an aesthetic achievement, however graceful and persuasive, must be translated into the ordinary life of the world. "My feelings about you now," says the letter-writer, "are those of anger and indignation, and they enable me to tell you a lot of things straight from the shoulder, without my usual tongue-tied round-aboutness." She amplifies: "You might as well take all your own literature and everyone else's and toss it into one of those big garbage trucks of the Sanitation Department, so long as the people with the top-cream minds and the 'finer' sensibilities use those minds and sensibilities not to make themselves more humane beings than the average person, but merely as means of ducking responsibility toward a better understanding of their fellow men, except theoretically—which doesn't mean a God damned thing."

This pointed outburst of scorn for certain academic values, this strain of anti-intellectualism, is not unusual in Williams' work. True, he distances himself from the remarks by attributing them to an anonymous, made-up character. But it is he, after all, who has fashioned the statement—as personal in its own way as another moment in *Paterson,* this one confessional rather than accusatory: "He was more concerned, much more concerned with detaching the label from a discarded mayonnaise jar, the glass jar in which some patient had brought a specimen for examination, than to examine and treat the twenty and more infants taking their turn from the outer office, their mothers tormented and jabbering. He'd stand in the alcove pretending to wash, the jar at the bottom of the sink well out of sight and, as the rod of the water came down, work with his fingernail in the splash at the edge of the colored label striving to loose the tightly glued paper. It must have been varnished over, he argued, to have it stick that way. One corner of it he'd got loose in spite of all and would get the rest presently: talking pleasantly the while and with great skill to the anxious parent."

With a novelist's eye for the precise details of an apparently meaningless moment in a busy doctor's life, Williams announces yet again a challenge that is both psychological and moral. Our preoccupations and obsessions, however innocuous and transient, indicate how hard it is for us to break out from what elsewhere in *Paterson* is called "the regularly ordered plateglass" of our thoughts. Williams approaches those small, daily conceits that keep many of us all too self-centered, conceits that confine us especially when the needs of strangers are at stake. For him plateglass is a contemporary medium, our version of that mirroring water into which Narcissus peered so eagerly and fixedly. I remember the old doctor waving his hands angrily as he looked at some of that postwar plateglass in New York City and thought about its capacity to separate people while promoting, through its transparency, the illusion of connectedness. Once, a bit enigmatically, he blurted out: "That glass is always painfully cold—the air-conditioning hits it in summer and the winter slaps it hard." He was letting me know cryptically how aloof and icy our solipsism can be. At times, as he waxed indignant, I thought of young Emerson writing his essays, such as "Self-Reliance," an earlier assault on the rigidities and smugness and arrogance of the universities—in Williams' phrase, "the whole din of fracturing thought." But he included himself in those denunciations. Like Emerson (in "The American Scholar" and elsewhere), he was interested in the distinction between character and intellect, and knew painfully that the former by no means necessarily correlates with the latter, and indeed, in some instances one finds a reverse (a perverse) correlation. As Williams once reminded me about the Nazi Joseph Goebbels and Williams' own friend Ezra Pound: "Look at the two of them, one a Ph.D. and smart as they come, and the other, one of the twentieth century's most original poets, also as brilliant as they come in certain ways—and they both end up peddling hate, front men for the worst scum the world has ever seen."

· · ·

If one impediment to moral conduct is the tug of our vain selves, another obstacle has been presented to us by another writing physician, Anton Chekhov. In the story "Gooseberries" we come across this remarkable, discursive statement uttered by the character Ivan Ivanych: "I saw a happy man, one whose cherished dream had so obviously come true, who had attained his goal in life, who had got what he wanted, who was satisfied with his lot and with himself. For some reason an element of sadness had always mingled with my thoughts of human happiness, and now at the sight of a happy man I was assailed by an oppressive feeling bordering on despair. It weighed on me particularly at night. A bed was made up for me in a room next to my brother's bedroom, and I could hear that he was wakeful, and that he would get up again and again, go to the plate of gooseberries and eat one after another. I said to myself: how many contented, happy people there really are! What an overwhelming force they are! Look at life: the insolence and idleness of the strong, the ignorance and brutishness of the weak, horrible poverty everywhere, overcrowding, degeneration, drunkenness, hypocrisy, lying—yet in all the houses and on all the streets there is peace and quiet; of the fifty thousand people who live in our town there is not one who would cry out, who would vent his indignation aloud. We see the people who go to market, eat by day, sleep by night, who babble nonsense, marry, grow old, good-naturedly drag their dead to the cemetery, but we do not see or hear those who suffer, and what is terrible in life goes on somewhere behind the scenes. Everything is peaceful and quiet and only mute statistics protest: so many people gone out of their minds, so many gallons of vodka drunk, so many children dead from malnutrition—and such a state of things is evidently necessary; obviously the happy man is at ease only because the unhappy ones bear their burdens in silence, and if there were not this silence, happiness would be impossible. It is a general hypnosis. Behind the door of every contented, happy man there ought to be someone standing with a little hammer and continually reminding him with a knock that there are unhappy people, that however happy he may be, life will sooner or later show him its claws, and trouble will come to him—illness, poverty, losses, and then no one will see or hear him, just as now he neither sees nor hears others. But there is no man with a hammer. The happy man lives at his ease, faintly fluttered by small daily cares, like an aspen in the wind—and all is well."

How do we find that "hammer" for ourselves? When those whom we love or respect (for example, our teachers) assert certain principles, urge them on us, hammer them home, we are likely to try going along. And our espousal can be deeply felt, no mere intellectual agreement. But Chekhov is getting at something else—our inclination, even when prodded, to respond only so far. His use of the adverb "continually" is singularly important—with it he makes it clear that an occasional knock on the head (a sermon, a lecture, or the reading of a story such as "Gooseberries") will not quite do. We shrug off, shake off, walk away from, close our eyes to the world of unhappiness. Chekhov notes the commonness of this maneuver: we stifle any inclination our conscience has to direct not only our awareness, but our conduct.

Not that we don't pay a price for our seeming happiness. In a devastating moral judgment, rendered almost like a whispered afterthought, Chekhov describes his "happy man" as "faintly fluttered" by "daily cares" and summons the image of "an aspen in the wind." Emerson might have hectored those in his audience with the exhortation to nourish the acorn of their conscience so that it might become, one day, a mighty oak, but Chekhov compares those of us who think of ourselves as reasonably content in life to the aspen. They can be arresting and lovely, banks and banks of aspens, especially in the early autumn, when they catch fire wonderfully before hibernating. But they assert themselves en masse, rather than individually. Chekhov steers clear, here, of an image that would suggest sturdy, willful independence, a willingness to go it alone. Aspens are a delight, but hardly the defiant loner. They blanket the land and give us a sense of familiar security. I do not picture William Tell standing in front of an aspen.

Chekhov can, at best, prick the conscience. We remember for a while that misery exists, but, made uncomfortable, we want to forget what we have just heard—and alas, soon enough we succeed in going about our business. Perhaps (and this is a gloomy thought) the best of us are in the tradition of Dr. Williams and Dr. Chekhov: we are seized by spasms of genuine moral awareness, but we are as pliant as aspens in our capacity to accommodate to the prevailing rhythms of the world we inhabit. It was an accommodation made by Williams, who knew (a painful recognition, he often said) that the poor and working-class people, his "ordinary folks," didn't read *Paterson*, even though he tried so hard to give them an important voice in it; and an accommodation also made by Chekhov, who knew well which people attended his plays, read his stories—certainly not those jailed on the Sakhalin Island he visited and described so painstakingly. The gnawing irony persists that powerful poems and poignant prose can affect us, excite us, cause us to see more clearly, yet not deliver that daily hammer-blow Chekhov prescribed.

Are we left, then, to savor irony, such a delicious part of the academic menu? Both Williams and Chekhov have their time with irony, the former in his fiery or truculent manner, the latter wryly, wistfully. The ironies of *Paterson* or those mentioned by Chekhov in "Gooseberries" unsettle us, though they are not exactly new discoveries. I have emphasized those ironies throughout this book, but I doubt that they can be repeated too often. That one can be well educated and not especially decent or kind-hearted is undeniable, obvious over the generations to all sorts of people from the highly literate to the relatively uneducated (some of whom may, in their "menial" work, perhaps on one or another campus, see ample evidence that big-shot intellectuals can be as petty as their so-called lessers). That one's happiness depends to a degree on a willful or unselfconscious disregard of the misery others constantly experience has also been evident over the generations to many of us who have been fortunate.

I remember my father talking at the dinner table about character, telling my brother and me, when we were young, that "character is how you behave when no one is looking." I also remember having a similar discussion with my psychoanalyst—and won't forget his response to my father's way of putting the matter: "He told you something that applies here, too [to the psychoanalytic situation]. For all the 'honesty' about emotions we encourage, there is a performer at work in all of us when we're analysands. We have to reckon with the analyst—and it's not all a question of transference." I wrote those words down long ago, and have mulled them over many times since then. Are we ever in a situation when "no one is looking"? In analysis we obviously must account for the doctor whom we are "seeing," and who, presumably, sees through all our dodges and deceits and is there to help us do likewise. But the billions of people who don't go through psychoanalysis (or read *Paterson* or "Gooseberries") nevertheless carry company inside themselves. In a sense, then, my father's wise remark has to be amplified: crooks and thieves do what they do because in their heads "no one is looking," apart from whether the police or anyone else happen to be nearby. For those of us who try to be conscientious, someone is always "looking," even if we are as solitary as Thoreau at Walden.

I suppose I would have my father's aphorism assert that character is how you behave in response to the company you keep, seen and unseen. The so-called psychopath or sociopath, the amoral one, is a person who has no such company, or maybe pretty bad company—the terrible silences of an emotionally abandoned early life or the demonic voices of a tormented childhood. In contrast, an overly conscientious person who hems and haws endlessly, if not crazily, about every possible matter is plagued by all too much company—strident and insistent commands that won't for a moment let go. Most of us, however, are neither criminals nor caught in the minute-by-minute tyranny of obsessions and compulsions. We live our lives in what gets called a reasonably "normal" way. Dr. Williams tells of us in his long, lyrical examination of America—the ordinary people who go about our business, don't get into trouble with the law, have our blind spots and, too, our reasonably good moments, when we are kind and thoughtful toward others. Dr. Chekhov tells of us as well—we whose happiness, he makes clear, is not necessarily misguided or a mean-spirited kind of selfishness, but rather often a matter of fate, chance, circumstance: the busy motions of the cheerful heart which leave little time or energy even for a consideration of the world's widespread wretchedness, let alone a concerted effort to take up arms against it.

· · ·

In this book I have presented the voices of students of mine who have taken to heart the moral analysis that Williams and Chekhov suggested. They have come to my office to talk, having done a good deal of quiet reflection on their own, and the result sometimes has been lively exchanges which prove hard to forget. Particularly memorable for me were the comments of one student named Gordon, who always asked a lot of himself: "I'm only twenty-one, so what do I know—I mean, about 'life.' I've been about as lucky as you can be. I've never really been in any trouble. All four of my grandparents are alive and healthy. My parents are happily married, and we've got plenty of money. All my friends used to say, when I was in high school, that I came from the 'ideal family.' I'm not saying everything was perfect. My dad has had a drinking problem; he's not an alcoholic, but when he gets depressed, once or twice a year, he binges out. He'll go at it for a few days, then level off—and we're fine for months. My mother gets very upset when that [drinking] happens; she'll be real down afterwards. She goes to church a lot. When things are normal, they don't go too much, except for Christmas and Easter. My mother has a bit of a money problem, I guess. She goes on binges herself, buying binges—like Dad, about twice a year. She builds up for one, and then—well, the bills start coming in. She went to see a doctor—I don't know if he was a psychiatrist—and he said she needed treatment. But she said she'd stop on her own. For two years she was fine. Then she had a buying spree, and she was going to start therapy, but she and my father talked, and they figured she'd spend more money seeing a doctor than she spent with all that plastic [the credit cards].

"But they're good people, and they've always tried to *be* good. My dad gives a lot of money to charities, and so does Mom. She inherited forty thousand dollars from an aunt of hers, and she gave the whole thing away. I was twelve or thirteen. I could tell that she had some second thoughts, and so did Dad, but they had one of their talks, and I was allowed to sit and listen, and I could speak up, if I wanted. I did! I suggested they give half to charity and keep half. My mom asked me why. I said—oh, I said, 'Because,' and I clammed up. My dad drew me out; he asked me what 'we' should do with the half we'd keep, if we stuck to my plan. I didn't know, at first. Then I remembered what he'd been discussing with Mom—that they buy a Boston Whaler [motorboat]. So I said we could do that, get the boat. Dad lowered his head, and then he said yes, we could. Mom said yes, we could, too. Then she gave her lecture on her aunt, and how she had been a volunteer at a hospital for twenty years, and she had raised a lot of money for the hospital, and she had given money to all these other causes, and we had to do what she wanted us to do. But she gave *us* the money, not the hospital—I said that. My father nodded; and I thought I was speaking for him then. His eyes and mine met, and I could tell! But my mother shook her head, and that's when Dad told her *she* should make the decision, and she did. All the money went to the hospital.

"I now realize that it was only a little while later that she went on another of her spending sprees. I mean, I now think there was a connection. I remember Dad saying to me and my little sister that he wished Mom had decided *not* to give that forty thousand away: he could use it 'now.' I still remember the look on his face. He seemed low. He *was* low; he said so—I heard him talking to his brother on the phone. That was my first lesson in psychology and in ethics. I had seen a real moral struggle taking place in my parents, and I began to realize that you can be high-minded and selfish at the same time, and sensitive to the needs of others and worried about your own needs at the same time. It was then that I realized my mother's buying wasn't totally a mystery; and it was around then that my father's drinking began to make a little sense to me—the way he'd come home from work and tell us that 'the company' (that's all I heard for years: 'the company') was putting 'pressure' on him and others, but he'd 'proven' himself and got his 'bonus.' The next day or so, he'd be coming home with slurred speech and stumbling all over, or we'd find him in the morning sleeping in his car, parked in front of the house. My mother would be upset with him, but I began to notice back then that she worried as much, maybe more, about the neighbors. What would they think?

"When I got here [to college] I thought I'd major in economics and maybe go on to business school. But I got bored with the numbers and the theories. I liked reading history, and I liked reading novels. I like working with computers, too. I'm my dad's son! At first I majored in history and literature, and then I switched to English. But I never liked the way the professors used the books—zeroing in on 'the text,' raking and raking, sifting and sifting it through narrower and narrower filters. I'm not against learning about symbols and images and metaphors, but there was something missing in those tutorials. Maybe I just had bad luck. There are some great people teaching here, but I didn't get them. All I knew—well, lots of snobby, 'literary' talk, lots of pretension and *phoniness:* that word Salinger used so much in *Catcher in the Rye.* They'd be so obscure about *Jude the Obscure* that you lost sight of the big picture—Hardy taking on Oxford and Cambridge and all the arrogance and privilege; Hardy taking on *us*, if you stop and think! You can read those 'texts' in such a way that they're not stories anymore; and the characters in them aren't people anymore, like you and me, all caught up in contradictions, and fighting to stay above water morally. But I'd better stop blaming them. It was me; I don't have an abstract interest in literature. I love to read stories and get lost in them, and some of the characters—they become buddies of mine, friends, people I think of. I hate to see a movie of a novel. The movie always disappoints me. I imagine the characters looking and talking one way, and suddenly it's different—there's Robert Redford playing Jay Gatsby. *Robert Redford!* I walked out after half an hour.

"I went back to computers. I ended up majoring in computer science, and that was when I started working in PBH [Phillips Brooks House, a student volunteer organization whose members tutor in poor neighborhoods and work with the elderly, prison inmates, and the homeless]. I went to Roosevelt Towers [a low-rent housing project] and I got to know some kids there, and they made my day. I'd play basketball with them, and help them with their schoolwork, and take them places in Boston and Cambridge they'd never seen; and I'd get them reading—not a lot, just enough so they could begin to get turned on. I'd read a story to them—one of Williams' about his patients, or even a Cheever story about the 'fancy richos,' one of the kids called them. I'd try to explain what was happening, but while I was doing it I'd suddenly remember some of those hot-shot graduate students with all their put-on airs. I'd cringe and hope I wasn't being patronizing. I hope I'm not turning those kids off. I don't think so. At first, they just put up with me. That's him, his thing—those books! But I think I got a few of them hooked. I love Salinger. I used some of his stories. His wonderful, electric anger, his contempt for people who think they're big shots and lord it over others, rang a bell with those kids. They listened. They asked me questions. We're on our way, I think, I hope, I pray.

"No, I don't go to church much. Lots of phoniness there.

Maybe I'm too hard on those ministers, and the professors. All I know is that you can go to church, or you can take a course in ethics and get high marks, and you can still be a pretty 'low-life' person—the way you behave. I learned that before I ever came here. My dad used to point out the churches, and he'd say they don't 'work.' He went because my mother wanted him to go. Once we drove by when people were coming out on Sunday, and he was pretty cynical. He used words like 'hypocrites' and 'two-timers,' getting a 'Sunday wash-job.' I think he'd started drinking. But you know, he wasn't saying anything different than Cheever says in some of his stories—for instance, 'The Housebreaker of Shady Hill.'

"When I have some big moral issue, some question to tackle, I think I try to remember what my folks have said, or I imagine them in my situation—or even more, these days, I think of Jude Fawley or Dr. Lydgate or Binx Bolling, or Levin in *Anna Karenina*, or Johnny Hake. Those folks, they're *people* for me. Nick Carraway or Jack Burden, they really speak to me—there's a lot of me in them, or vice versa. I don't know how to put it, but they're voices, and they help me make choices. I hope when I decide 'the big ones' they'll be in there pitching."

The moral contradictions and inconsistencies in our personal lives more than resonate with those in our social order, our nation's politics, our culture. As my students keep reminding me, one can go to the Bible itself and find plenty of those same incongruities, those clashes of different or opposing values, ideals. Nor have our universities been all that successful in figuring out for themselves what their obligations are with respect to the moral questions that many students put to themselves. In that regard, the young man just quoted at some length could be quite unsettling: "I've tried to take courses in moral philosophy. I read the books. I become smarter in the analysis I do. But I leave the lecture hall and I can see myself as the same—the way I'll think of certain people, the way I'll behave. I guess this place is where your *intellect* changes. I guess it's no behavior-modification place. But a history tutor said that Harvard used to be a place where they worried as much about the students' morality, their character, as they did about how well they memorized books and wrote exams. That was in the eighteenth and nineteenth centuries. Not now: we're way 'beyond' that here. It's each person to himself on a lot of these moral issues, so long as you don't break any laws or rules. One day that's fine with me, but the next I wonder whether that kind of attitude will be enough for me when I get married and have a family."

Meanwhile, Gordon tries each day to live his life, to get those "good" grades that will advance him along, give him one or another boost; and he tries, rather often, to *be* "good," to live up to the notions of what that "good" is which he brought with him to college, and to modify or expand (or weed out, hone down) some of those notions in accordance with the courses he's taken, the experiences he's had with his roommates, his friends, his teammates (lacrosse), and, not least, those he has tutored through Phillips Brooks House. He keeps telling himself, as he said during our talks, that he counts among his "friends" characters in certain novels. He denies any distinctiveness or originality in that kind of friendship: many of his high school buddies, he reminded me, "would talk about Holden [Caulfield] as if he was one of us." Having said that, he asked me a rhetorical question, well worth a pedagogical conference or two, I think: "Why don't you guys [college professors] teach that way?" What way, I wanted to know. "As if Holden was—I mean *is*—as real as you or me." I fear my response was all too defensively, protectively, ingratiatingly self-serving. But he wasn't really interested in an *ad hominem* discussion or in an argument. He was reminding me, really, of the wonderful mimetic power a novel or a story can have—its capacity to work its way well into one's thinking life, yes, but also one's reveries or idle thoughts, even one's moods and dreams. "I come back from working with those kids and I think of what Chekhov wrote in 'Gooseberries.' What he wrote isn't just another paragraph in a story; it's what's happening to you, right here, right now."

So it goes, this immediacy that a story can possess, as it connects so persuasively with human experience. Dr. Williams and Dr. Chekhov and the needy children whom Gordon got to know as his tutees can offer their own kind of moral instruction. Dr. Williams urges intense, searching self-scrutiny. Dr. Chekhov urges a close look not only at ourselves but at others, at the terrible contrasts of this world. The tutees urge—well, themselves as fellow human beings who have something to give (their neediness becomes someone's opportunity to be generous) as well as receive (the instruction they get). All in all, not a bad start for someone trying to find a good way to live this life: a person's moral conduct responding to the moral imagination of writers and the moral imperative of fellow human beings in need.

Index